COUNTRY VET

Hope you Enjoy it
Daniel S. Dennis

COUNTRY VET

Danniel S. Dennis

NPI

Northwest Publishing, Inc.
Salt Lake City, Utah

NPI

Country Vet

For information address: Northwest Publishing, Inc.
6906 South 300 West, Salt Lake City, Utah 84047
JC 09 30 94
Edited by: R. Larsen

PRINTING HISTORY
First Printing 1995

ISBN: 1-56901-234-2

NPI books are published by Northwest Publishing, Incorporated,
6906 South 300 West, Salt Lake City, Utah 84047.
The name "NPI" and the "NPI" logo are trademarks belonging to
Northwest Publishing, Incorporated.

PRINTED IN THE UNITED STATES OF AMERICA.
10 9 8 7 6 5 4 3 2 1

Table of Contents

Preface

To write about your life is a humbling experience, for you realize that it will expose many inner thoughts and feelings for all to see. It's far more comfortable to keep them shrouded in your mind, to share in limited parcels with those you choose. Family and friends encouraged me to write this account. Without their repeated support and willingness to help I would never have made the attempt. As I have looked at this memorable journey through life, it has rewarded me with untold joy and appreciation for all those who have shared it with me.

Speaking of those who share in these few episodes, I wish to express my heartfelt thanks for having had the privilege of traveling the road with them. One disappointment has been that I could not recall many of the names and therefore I've had to leave them out. In a few instances names have been changed to avoid embarrassment. I hope that none will feel any criticism beyond that which I feel in laying our lives open for others to see.

My effort as a pioneer in veterinary medicine for this great Uintah Basin area has been most fulfilling and satisfying. Let it not be said that it didn't test the very metal I was made of.

The challenges and disappointments seemed to come with the territory, and each has been a growing experience. I hope the reader can share in the triumphs of overcoming them.

In thirty-five years of practice, literally thousands of people and their animals have crossed my path. I've tried to pick out a few of those that stood out as unusual, amusing and very typical in my memory. I've tried to write and record them in common everyday terms for all to understand. Some things are very candid and may offend some readers and for this I apologize and beg their forgiveness.

I wish to give thanks to all those whose lives have touched my own and a special thanks to my family and friends, without whose unfailing encouragement this would never have been finished.

1
THE DAWN OF HOPE

The incessant yapping and howling of the coyotes just up the ridge drifted in and out of our dreams until they finally woke us up. We couldn't get back to sleep so we decided to get up and start the day.

There was a chill in the air that made my clothes feel good as I slipped my legs into my pants and stood up on the hard bed. Last night when we scattered the straw and threw out the tarpaulin and sleeping bags it felt comfortable, but now it seemed hard and had a few jabbing lumps where rocks prodded us to move on. It was a dark night and the moon was down leaving the stars vivid and bright with the Milky Way like a belt holding it all together. I looked to the east for some hint of the coming sun but the sky was all the same.

When the fire began to flame, its light faded the stars somewhat and I had to step out past the half dozen pinion and cedar trees to check the eastern horizon. It seemed a little bit lighter and the brightness of the stars had faded slightly as I gazed up trying to comprehend that there was no end to that space out there. It just went on forever.

Later while the horses munched the last of their oats, I checked the sky again to see a little green hue with the stars practically gone in that direction. The faint shafts of pink light sat on the mountains about a hundred miles across the Basin.

What was formerly shadow became the magic of a new day being born and it caught and held my attention as if the changing drama was a giant magnet. The light pink hue in the now blue green, gradually flared into a fan covering the eastern half of the sky. At the center, a shaft of brightness showed with a touch of yellow. As the yellowish brightness increased, the just perceptible pink moved quickly westward along the north and south horizons to join a deep blue shadow and the green faded to leave varying shades of blue overhead. Then the brightness of the small yellow fan intensified eastward until a shaft of white-hot light came over the edge of the horizon like molten iron. It seemed to vibrate and swirl with motion as it widened with a rounded top and grew into a half circle. At this point its intensity began to force my eyes away and as I beheld it in relation to the visible horizon it was a huge seething ball of white-hot light that appeared perhaps seven times the size of what we see overhead at noon. The morning was born. We could follow tracks and it was time to gather cattle.

Like a lot of young people moving through high school, I was groping in the dark for something to become, something to devote my life to, a goal to reach for.

I wasn't very good at sports and was too bashful to get involved in speech, drama, or even dating and social activities. I knew how, and loved to work, had a knack for driving and handling machinery, and a love of the soil. I guess most of all I had a Mom and Dad who repeatedly encouraged me to get an education and not be like them, having to work at whatever they could and farm. They wanted us kids to make something of ourselves and have a little easier life. Up to this time, school had been just something everyone did—you had to because it was required. I was sliding along just waiting until I reached eighteen when I could quit school and go drive a truck or something. As I entered high school I began studying about agriculture and got involved in the Future Farmers of America program. For the first time I began to study because I liked it. I did very well and made the crop judging team and won in public speaking. We had our own chapter house with some library offerings and I began to read, especially as we traveled to and from school from Myton to Roosevelt.

Every boy had a home project, most of which were calves which each boy fed and showed. I think I would have liked a calf, but our few head of cattle were like a lot of others, in a state of transition. Dad had started with Old Whitey, a roan Durham dual-purpose type on which we all learned to milk at age five or six. As we acquired more we moved more toward the dairy type because we were milking five or six for the weekly cream check. A dollar or two in cash was hard to come by in those times. We also had a neighbor with pure-bred Jerseys and we used his bull in the early years. Then times changed and Dad was forced to work away a lot in the mines and building canals. This started us in a different direction, toward raising cattle for beef so we wouldn't have so many cows to milk. The result was a small herd in between, without a good enough quality heifer for a dairy project and no steers to measure up in the beef field. Buying a calf for a project wasn't considered because our farm mortgager was threatening foreclosure and keeping up with the taxes strapped us financially all the time.

I think Mr. Atwood, my Ag Teacher, understood the situation and made a suggestion that I consider pigs. We always kept an old sow to furnish us with some pork and sold the neighbors a pig or two for their meat. Transportation was improving so pigs could be fattened out and shipped to market from this area and I think he was trying to push this as another phase of agriculture. A lot of people like us had a pig for local consumption and the feeding practices more or less revolved around what was available. Some had milk to go with the slop from the house and plenty of hay or pasture and a limited amount of grain. The quality of the pigs wasn't bad, since everyone kept it in mind when selecting breeding stock, but there were no pure-bred breeders that I knew of in the Basin. I think my teacher had this in mind when he suggested that I purchase a young pure-bred boar and a gilt to start a pure-bred pig operation. He said if I bought them young they wouldn't cost too much and I could grow into the business of selling breeding stock. He felt I might do a little better selling stock than feeding pigs for market.

I went to the Roosevelt State Bank with my dad and laid out my plans before Mr. Ray Jordan. He was willing to loan me the fifty dollars I would need. I'll never forget the lesson on debt that he taught me that day. When Mrs. Orser brought the typed-out note and laid it before me and I signed my name, he looked me in the eye with a little smile and said. "Now, Dan, doesn't that make you a little sick to your stomach?" It took a moment for the meaning to sink in and I nodded my agreement. I was feeling quite elated that I could borrow money, just on my signature, but when he said that it sure changed to humbleness.

I went out to the Heber valley with Mr. Atwood on a very cold day just after Christmas and purchased the pick from two litters of pure-bred Duroc Jersey pigs. They were a small feeder size and we brought them home in a crate in his trunk. I was launched on a four-year project at age fifteen.

Mr. Atwood was a great teacher and used experiences as teaching tools. He did get some boys to start pig feeding projects and one boy's project served as an unforgettable lesson.

The boy's parents were of mid-European origin, probably from Austria, and they had a litter of pigs which he took as his project. When Mr. Atwood checked his project he found the pigs were seven months old and not yet fat enough for market. He went to great lengths to explain that pigs could be fed out in about five months if they were given grain free choice. He gave the boy and his father plans for a self feeder and explained the benefits of two litters a year and chastised them a bit for theirs being as old as they were. He knew the father was the problem and talked mostly to him. When the old man felt like he'd heard enough he took in a short breath and this was his comment. "Aw vaut da hell is time to a pig anyvay?" We remembered the lesson.

The high school years were a growing time in my life when I began to search for a future. Our farm was small and deeply in debt and I realized it wouldn't sustain any more than just Mom and Dad. Other employment opportunities in the Basin were almost non-existent. What should I try to do? Where could I make a living? If I did go to college what should I study? There were all these questions, and the fun of intellectual achievement which had just begun was suddenly very clouded and uncertain. I was groping in the dark, mostly because I didn't know what was out there.

High school graduation hadn't been one of my highly anticipated events, but as the time grew nearer I seemed to sense that it was a special achievement, a milestone that would be a new beginning. I was interested in the plans, and not unwilling to take time off for the practice sessions in the LDS Stake House where it was to take place. The graduates were all to be seated in the circular rows of choir seats behind the speakers at the podium. Upon the proper signal we were to arise in unison and file out row by row to individually pause on a small platform built out from the podium as our names and that of our parents were announced. We were then to descend the three small steps to get our diploma and take our place in the first rows of the chapel. With a couple of practice runs we soon had the routine down and all was ready, except for the embarrassment I felt of having to do it in my corduroy pants and the well-worn blazer I wore to school. I hadn't bothered to go, or even let myself think about going, to the

Junior Prom or the Senior Hop because I didn't have anything suitable to wear and we didn't have a car. I rationalized that they were too "highfalutin'" for me, but now it was graduation. I wondered how I'd feel. How proud could I be to step into the limelight with everybody looking at me in those common, everyday clothes? Perhaps this feeling was transmitted to Uncle George when I happened to meet him on the street in front of the bank.

"So, tomorrow you graduate?" was his happy greeting.

"Yeah, I guess so," I said a bit dejected. There must have been some inspiration concerning why I felt as I did.

"Have you got a suit, young man?" He already knew that I probably didn't and I knew he knew that.

"Nope," I said as I shook my head.

"Well you come with me into Penney's." It was just a door down the street and off we marched. My heart began pounding with excitement and my step quickened. The thought that I could have a suit made the sunlight of my life burn a lot brighter and I was rejoicing inside. They picked out several suits about my size and each one felt good as I tried them on. I was so thrilled it didn't matter which one to me. When I tried on the last one it was a dark green hard-finish wool, a little bigger, free fitting and comfortable. I couldn't believe it was me in those mirrors.

"That's the one," Uncle George said with a big smile on his face. I was having a hard time with reality. I really was going to have a suit to graduate in and probably no words can ever express how much it meant to me. No other suit I've ever owned has meant more to me than my new green suit for graduation.

It was with great pride and anticipation that I took my place on the stand that graduation night. I listened with squared shoulders and head erect to the speeches and rose with pride as we all stood up—the girls in their beautiful long formals and we boys in our suits. The graduates in the front rows started moving forward with the girls' formals trailing behind and more than a few of the boys following stepped on the dress of the girl ahead momentarily. I was glad I noticed so I wouldn't make that mistake. One by one each graduate was announced, a girl then a boy, and another girl.

I followed Leona Shields and there were only a handful of kids behind us. I worked my way around the semicircle and down to the platform and finally it was my turn. I stood tall with my shoulders squared, my head up, looking right into the eyes of everyone seated in the audience and stepped forward onto the platform just like they told me to. How proud I was as I stood there in my new green suit with the padded shoulders that made me look so grown up, strong and confident. How good my folks must have felt to see me as their names were read. No other event in life had ever made me feel so uplifted as this moment did.

Then, as I glanced down at the steps before stepping down, I beheld Leona's beautiful formal spread out over the platform with my two feet planted firmly on it and her leaning into the dress, like a horse in the harness, waiting patiently for me to wake up. If ever one could descend from the epitome of pride and exhilaration to the depths of embarrassment and shame it happened to me in those few seconds. In all the years since then it has always remained my life's most embarrassing moment.

I had graduated, but our Future Farmer projects carried on over into the next summer. Most of these projects were livestock and during the high school years I had been aware that some animals had experienced health problems. Our teacher could handle most of them, but occasionally it was beyond his capabilities. We felt a need for help which was not available in the Uintah Basin. This need, like the dawning of a new day, became how veterinary medicine was born in my mind.

If I could somehow go to school and become a veterinarian, I could bring the service back here to the Basin where it was needed.

My dream evolved and burned with the intensity of the sunrise in the morning. It was a good thing, because the intervening years saw many roadblocks and rough places, with severe tests before its completion.

The first problem seemed to be the lack of opportunity to go to college. Our farm, like most of the others in the Basin following the great depression, was in debt. My father was just getting out of a full body cast from three broken vertebrae acquired in a mine accident seven months earlier. My older brother Howard was attending his second year at Brigham Young University and was struggling to make his own way. I just didn't think it was possible for me to go unless I could get a job and save some money. I doubted that I would ever be able to go.

In the early summer, just after school was out, my Ag teacher, Mr Atwood, came out to make his final checkup on my swine project. He asked me where I'd be going to college and if I was still interested in veterinary medicine, since I'd shared my dream with him. I told him I didn't think I would be able to go at all and explained about the farm and our family situation. I'll never forget how stern and intent he was as he looked into my eyes with a gentle sincerity and a feeling of love that penetrated my very soul. Pointing his finger at me he said, "Dan, you don't need money to go to school, all you need is a little intestinal fortitude. Now you think it over. I think I can get you a scholarship for your first quarter's tuition and you can make it from there if you want to."

I did think about it; in fact, I couldn't get it out of my mind. When we finished stacking the first crop of hay I put on my best and only pair of corduroy dress pants and a clean shirt and hiked down to the highway to try my luck at hitch-hiking.

It was a beautiful, clear, summer morning and I guess that helped because Logan, 240 miles away, seemed very far. Not knowing anything about Utah State Agricultural College made me a little frightened and very apprehensive. I'd never been further than Salt Lake City and I'd only been there two times, but I'd looked at a road map and knew the highways I needed to take. I felt I was calling upon that intestinal fortitude about as much as I could and had no idea how I was going to approach the people at Utah State once I got there. I think the Lord was on my side because I wasn't standing there very long before a car picked me up to further confirm my commitment.

With my uncertainty, I didn't share my purpose with my benefactors unless pressed and then my answer was just to check out Utah State. The rides came easy and I was soon in Salt Lake. I knew the road to Logan went north from Salt Lake but it was a big city and a whale of a long way to walk just to get out on the highway again. I started to hike up the street to the north and was just passing a bus stop when I noticed the second bus said North Salt Lake. "Hey, that will put me out in the direction I need to go," I thought, so I climbed on and paid my nickel or dime fare (I don't remember which) but it wasn't much. I just figured I'd ride to the end of the line then find the highway from there. Things were going great and I kept watching for the highway signs and boy wouldn't you know it, we were on the highway I needed, 91 and 89 both. My luck was coming up good at every turn and it wasn't long until I was at Brigham City. As I hiked toward the canyon with the afternoon sun behind me I was watching my small shadow move up the pavement in front of me. I heard a car and turned with my thumb out.

Two young strangers in an open Model A Ford coupe slammed it into second gear and got it stopped and waved for me to get in the rumble seat. They were a couple of what might be termed "the carefree type" on their way to southern Idaho.

I knew my destination of Logan was assured, but the way they careened around those canyon turns made me wonder if we were going to make it at all. The old road from the summit of Sardine Canyon went east toward Hyrum down the mountain, then swung back westerly toward Wellsville across the foothills. As we came up over one of those rises, doing all the Model A would do, there in the road from fence to fence was a big herd of sheep. The driver jammed into second gear and stomped on the brake pedal while his companion hauled back on the emergency parking brake with all his might, but I knew from previous times when they'd tried to slow it on the curves that it wasn't going to work.

We slammed into those sheep and bounced to a stop, right side up, out in the middle of the herd. After the dust cleared and the herd moved on there were only three that couldn't limp off. The herder was busy cutting their throats to bleed them out for meat and my friends were checking the car to see if the wheels were on and the tires still up. Then came the well punctuated lecture from the owner of the sheep and the offer to pay by the thoroughly humbled and frightened boys.

As they emptied their pockets and came up with just over ten dollars I was fighting a battle within myself, wanting to help, but knowing I was a long way from home with a night to spend. I had one five-dollar bill in my shoe and a little change in my pocket. I finally did pull out the change to offer them, but they wouldn't take any from me. They said it was no fault of mine. They gave the owner all they had and offered to send him more from home, but he, having calmed down some, mentioned that he'd be able to use the mutton and that was enough. We were on our way again and in tempered silence we soon covered the miles to downtown Logan.

I picked out Old Main at the head of 500 North as my guiding beacon and began hiking toward my target. By the time I reached campus and headed up the walk and stairs my intestinal fortitude was about used up. I didn't see the grass, the flowers, the trees and the beauty of the setting. It was just some inner power that kept my legs climbing onward. I was all alone. No one passed me, no one was in sight. Homesickness welled up inside me and I felt far away from home, small and insignificant. It had been a long day and maybe the sheep incident taxed my nerve. I wanted to turn and run toward home as fast as I could go, but somehow I kept my head.

When I reached the front door it didn't get any easier. I had to stop and regroup my scattered courage and determination once again. I looked at the doors, which were taller than usual and had tall windows in the upper half set in at least three inches of thick dark hardwood. I didn't have any idea who or what I was going to find inside or what I was going to say when

I did find someone. I pulled open the door on the right and stepped through. It was about 4:00 P.M. and somewhat darker inside with no lights on in the hall. It took a moment or two for my eyes to adjust, then I saw "President's Office" on the first door on the left and it was closed. Somehow that gave me a little relief. I noticed that all of the doors down that huge hall were closed except the last one on the left. There was light coming from it, so I started walking that direction. I wasn't hurrying or stamping or anything but my steps echoed down that hall like a shod horse and that hall seemed as long as a row of corn to hoe.

When I finally came near the opened door I could see a girl sitting behind a desk and a wave of relief swept over me. There really was someone there and she didn't represent the image of the college president my mind had invented. I kind of stopped part way in the door and she looked up, "Can I help you? Come in."

"Well, I ah…I guess so." My mind raced for an easy way to tell her why I was there but nothing came. "I want to go to college and I don't have any money." I hadn't minced any words; I'd made my confession and for a moment she seemed a bit puzzled as to what to do, like "who should I send this kid to?" Then she smiled a little with understanding as she read my mind.

"You want a job?"

"Yeah, that's what I wanted to check on, my Ag teacher thinks he can get me my first quarter's tuition on a scholarship and if I could find a job I might be able to go to college." I'd spilled the whole story, short and to the point. I was impressed that she was very efficient, because she pulled out some kind of form and laid it in front of her.

"What's your name?"

"Dan Dennis"

"D-E-N-N-I-S, right?" She was writing it down. I thought, "This is for real, it's going to be official, I better give her my full name."

"Yeah, that's right, and my real first name is Danniel."

"D-A-N-I-E-L?"

"No it's got two N's in it."

"That's different. What's your middle name?"

"Stewart. My mother was a Stewart."

"We'll just use the middle initial. Where are you from?"

"Myton, Utah."

"M-I-T, how do you spell that? I've never heard of that town before. It must be down in southern Utah." Saying she had never heard of Myton diminished my ego another notch, but already I was feeling more relieved.

"It's M-Y-T-O-N; and it's out in the Uintah Basin about a hundred and fifty miles east of Salt Lake."

"Oh yes, out toward Vernal and Craig, Colorado. What high school did you go to?"

"Roosevelt High."

"That's a long ways; have you come all that way today?" I nodded that I had, then she said, "Are your folks out there? Do you want to go invite them in?"

"No, I just came by myself. I hitch-hiked."

"You mean you've hitch-hiked all the way up here today. I don't know how you made it."

"Well, I guess I was pretty lucky."

"What do you want to study when you enter college?"

"I want to be a veterinarian. I guess I just want to take the classes it takes to get into vet school."

"We'll put down agriculture as your major and write pre-vet in where it says minor, that's what most of the pre-vet students do."

"Whatever it takes. I don't know too much about it yet."

"Let's see, what kind of work have you done?"

"Mostly farm work. My folks have a farm. It's only 80 acres, but it has plenty of work—then I got a job driving tractor once and I've worked in the timber a little. I've helped mix cement for structures in the canal, and I can use a hammer and an axe about as good as anybody. If I could get a job, I think I could learn how to do about anything and I really like to work." I felt like maybe I was getting a little carried away, so

I stopped trying to think of things I'd done.

It was about this time when a gentleman stepped out of the back office and came over to me. It looked like he was wearing a suit, but he had his coat off and he was kind of smiling. He offered his hand and I took a good hold and shook it a little.

"My name is Les Pocock, I've been listening. Have you ever milked cows?"

"Sure have, since I was about six years old." He held on to my hand then put his other arm across my back and around my shoulders, then in a loving fatherly voice he said.

"Son, if you'll just show up here this fall when school starts, I promise you I'll have a job for you. There isn't much going on this summer but you come this fall and I'm sure there will be a job." He seemed just like Dad to me and I got a lump in my throat for a minute when I was trying to thank him and the girl. My spirits were soaring like never before as I bid them goodbye and as I hurried up the hall I felt like I was walking up in the air. When I stepped out through those same doors it was a different world somehow, there were flowers and trees and beautiful lawns. I'd never seen a more beautiful place, the whole valley was green and there were no brown hills like we had at home.

By the time I'd walked back down to Main street in Logan I had decided I'd better spend the night. I looked at the main hotel but passed it up and went around the corner to a smaller one. The man charged me two dollars and fifty cents for the night. I bought a hamburger and fries and I even went to a thirty-five cent movie. I was really living it up and felt like the world was on my side. Early the next morning I started for home.

That fall when I returned to Utah State, Mr. Pocock gave me a warm welcome and asked me if I'd mind milking the college dairy herd at 4:30 A.M. I was glad for the job and it turned out to be a great blessing. It got me up in the morning to a good start and my job was over before classes started. I could drink all of the cold milk I wanted there on the job and with about a quart as my breakfast each morning I could make it through the day until our evening meal. It wasn't long until I was working in the vet science lab at night to add a few more of those "thirty-five cent hours" to help out.

Seven years later, with World War II included, I completed the first hurdle—graduating from Utah State and being accepted at three veterinary schools. In the process, several other major threats thrust themselves to the forefront to challenge and almost extinguish the light of my original dream.

2
WAR AND DECISION

In 1939 Hitler marched into and conquered Poland and by the time spring quarter was to start in 1941 there was no question in my mind that the U.S. would soon be in the war. With this inevitable war on the horizon, an opportunity to prepare myself came along in the form of a civilian pilot training program at Utah State. I raked up the ten dollars, took the physical and enrolled. It was a major change from pre-vet classes to meteorology, aerodynamics, navigation, aircraft engines, morse code, and flight training.

I finished the first course and received my private pilot's license. I then went on into acrobatics in the secondary program and by then I was completely hooked. Flying was in my blood and without a doubt there was nothing that could compare to it. My future and my life revolved around flying, and veterinary medicine was a dream of the past. I had to spend a week or so after school let out in the spring to finish my courses, but who cared? I was having fun at something that seemed my second nature.

Then came Pearl Harbor and my enlistment in what was then known as the Army Air Corps. Because training facilities were not available I wasn't called to active duty right away, so I remained in college and expanded my military preparation.

When I was called up I went to Sheppard Field at Wichita Falls, Texas for my basic training. From there I was assigned to Oklahoma A & M College at Stillwater, Oklahoma for more college preparation. As the training facilities expanded, I completed my pre-flight training at the cadet center, now known as Lackland Air Force Base at San Antonio Texas. Because I had previous flight training, I was assigned to train with the Royal Air Force Cadets from England who were being trained here in the States. I went to Terrel, Texas, just east of Dallas, where I completed all of my flight training. I received my wings and a commission from the U.S. Air Force and earned my wings from the Royal Air Force as well.

Because the training received with the R.A.F. included extensive navigation, low level, and night navigation, I was assigned to the Air Transport Command. My first assignment was to Romulus Air Force Base outside of Detroit and our mission was ferrying planes to the war fronts. I checked out in and learned to fly as many planes as time permitted, and took advantage of a lot of simulated instrument flying time in the link trainers available there.

After about two months I received orders to go to India where I logged over a thousand hours flying cargo and passengers across India and over the hump into Burma. I loved flying, and discounting the many split-second reprieves from suddenly exiting this life, it gave me excellent training and experience to fly the airlines after the war. I finished my required one thousand hours at about the same time the war ended, but my return to the States was delayed for a few months while I waited for transportation back home. I came back with great hope, planning to enter the airline service. When I applied to my first airline there was a waiting list of over 4,000 names. It was disappointing to be in the late group getting home, but I remained undaunted, thinking my experience in weather, instruments, and passengers would sift me to the top. I just had to wait until it happened.

In the meantime, with G.I. schooling money available, I decided to get back into college and get a degree while I was waiting for the airlines. With so many credit hours already earned toward an agriculture degree, I resumed my former goal. Periodically I touched base with the airlines. The list was diminishing and they encouraged me to get my commercial pilot's license and check out in comparable planes. When I started checking around and found it would cost me around one hundred and fifty dollars per hour to do this, it was financially impossible.

This dampened the flame of flying airlines as a career and the practice of veterinary medicine was getting closer to my grasp. You had to apply to the vet schools six months to a year ahead which I did in 1946, not having quite all my requirements completed. As each letter came from the schools informing me that I wasn't accepted, the little flame from long ago burned lower, but I was encouraged and counseled to try again.

Great turmoil was gripping me in my love life and deep questions in the spiritual arena were awakening me to a greater understanding of the gospel. The deep love that I felt for my wartime sweetheart was dampened by unresolved guilt feelings that I wasn't worthy of her. Not being able to see her, together with the realization of the differences in our lifestyles, plagued my thoughts. I was financially rather poor and she was used to more affluence. I knew that if I were to fly I could probably meet her expectations, but if I were a struggling veterinary student or a dirt farmer I wasn't sure she could cope with it. I was also deeply disappointed to come back from the war to find that she had started smoking. She said she could quit and maybe she could have, but having been a non-smoker and forced to clean up cigarette butts from others all through my military career, probably had an undue influence upon me. As the potential for a flying career gradually faded I forced myself more and more to question our future in spite of my deep feelings.

At the same time I had been dating other girls a little and had met one that began to kindle another fire. Joyce was beautiful, a little younger, came from a financially struggling family and was putting herself through school by working. She had all of the important skills that a wife should have in my way of thinking. She could cook and sew and came from a family where love was evident. She was liked and respected by her peers and was a model of obedience to the standards of the gospel. Since I was in the process of coming into the Church of Jesus Christ of Latter-day Saints, she was a choice example that I knew would be good for me. I met and liked her family and I felt this was important. All these factors came into play as I continued my studies. I continued to date and court Joyce, and her presence and response twirled the knobs of my heart until the tumblers rolled over and I found myself in love with her.

While this major change was occurring in my life, the hope of flying dimmed even more. It was time to apply again to the vet schools and this time I was accepted at three of them. The anguishing internal debate time was over. I had to make a decision. I didn't have the finances to sharpen my skills and acquire the commercial license for a nebulous chance to fly, but I could go to school on the G.I. bill. My heart seemed to tear a little as my mind made the resolve that my first love and flying would have to remain cherished memories. I rationalized that veterinary medicine was needed back home by a lot of good people, and that my desire to fly was sort of a personal fling of my own. It helped to rationalize this way, but the actual pain of letting the dream go took a long time going away. It was one of those burdens you have to carry alone because no one else could quite understand.

It was easy for me to decide I wanted to go to Colorado A&M because it was my first choice and closest to my home. I wrote the letter of acceptance and also the thank-you letters to the other two schools. I was committed to another four years of college and I knew they were not going to be easy. I knew that Joyce was young and not yet ready to get married, but I didn't want to leave her and go to Colorado alone. I felt that if I did I might loose yet another love and I was halfway through my twenties.

The year and a half after leaving the Air Force was truly a period of turmoil and great decisions, which taxed my mind and my time. I was working part time, carrying a full load of classes and trying to win my new love's hand. The most amazing thing of all was that I made better grades than any other time in my schooling. I guess it was a vague bit of proof that we have much greater mental capacity than we normally use. At any rate I did graduate and we were soon engaged. The summer was filled with intense preparation, a few short but loving visits, and ended with our marriage in the Salt Lake Temple on the tenth of September. During my visits, my future in-laws encouraged me to look at going to medical school since it entailed about the same amount of time. I considered it, but the influence of growing up in the out-of-doors close to the soil outweighed the possibility of spending all my time in an office or hospital. Once again the scales tipped to the animal side and the dream of long ago flickered brighter.

As newlyweds of ten days or so we traveled eastward through the beautiful fall colors of the Colorado mountains to our tiny apartment at 1020 Akin Street in Fort Collins, Colorado. It was our first home and the joy and happiness of this great adventure has lived through the years to repeat itself in many more trips over those same miles with renewed joy and deeper memories. Life for me had never been sweeter or filled with more joy and although Joyce had a good measure of apprehension and homesickness she was like Ruth in the Bible, a hundred percent committed and willing to do her full share. Happily, there were two sets of parents equally committed to give us their full measure of support and love.

When I walked into my first class, which was anatomy, and took a seat in the middle to be just one of the group, my apprehension was running mighty high. I'd learned that this was the class where the weeding out would take place and I hoped that I wouldn't be one of the unlucky ones. There were sixty of us and all were veterans of World War II except two. I knew they were all serious and men of experience. The professor held up the text we were to use and explained that in the course of our year and a half of study we would literally memorize what was contained in that book. It was a large book and contained over 950 pages. My heart really sank. I had never been very good at memorization and such a challenge as this seemed impossible for me. For a time I was wondering if I really had what it was going to take to become a veterinarian, but I looked around at the others in the class and could see they had the same concern that I had. I decided that I wasn't going to throw in the towel, they'd have to kick me out. I calculated that it was just a threat, a figure of speech, to let us know it wasn't going to be easy, but the odd thing was that in the course of the next two years we did literally learn all that was in the book. Each bone, attachment, muscle, tendon, vessel, and nerve came alive in our memories as we dissected and held them in our hands or saw them under the microscope. When I took the first test, I realized that what was written in that book was the most logical, organized way of describing each part of the animal body, and from that time on I placed in my memory a pattern of how a part was described and by using that pattern and my own words, from the pictures in my mind I could very closely repeat what was in the book. I ended up among the top in the class.

It wasn't easy to cover all we had to learn about the various species of animals, and our days were long. We listened to lectures and did laboratory work from 8:00 a.m. to 5:00 p.m. Then we studied from about 6:30 p.m. to midnight every night, seven days a week. We were assigned a partner to work with in the dissecting laboratory and we usually studied together during the supper-to-midnight effort. The role of the wives was equally challenging. We spent only the mealtime together. The wives sat quietly in a corner knitting or sewing to pass the time while we quizzed and rehearsed each other on the subjects of the day. Weeks passed into months and months into years. The pace made it all move fast as we packed our brains with much needed knowledge.

Having been released from the service with an automatic reserve status for the next five years, another threat presented itself. The Communists thought they could force the U.S. out of Berlin by a land blockade, but Uncle Sam wasn't about to give up. The famous Berlin airlift got underway and since my military occupational specialty was cargo pilot, I received several letters inviting me to return to active duty to fly. I was in my junior year with a wife, a baby girl and expecting another and they could not go with me because it would be a temporary duty assignment. The thought of flying again was a real temptation, but when I looked at my little family and knew I would have to be away from them, I soon realized that the right thing for me to do was stay in school if I could. I also knew they could send me a set of orders calling me back to fly if they chose to do so and the urgency for pilots seemed to be building. At this particular time an opportunity came along to accept a commission as a veterinarian in the military as the result of participating in the Veterinary R.O.T.C., but I would have to serve two years upon graduation. Another dilemma presented itself—I could take a chance I wouldn't be called up to fly and finish school, or accept the commission, get paid for finishing school and serve the two years. My duty seemed to say, "stay with your wife," my love of flying and adventure said "take the chance," and my real self was struggling with what to do.

Sometimes financial considerations have a way of overriding all other choices, and that was the case with us. Our G.I. Bill money was all used up. I had worked and saved during the summer but we were to the point of borrowing to stay in school. The second lieutenant's pay would take away the financial problems and the two years duty would be inside the continental U.S.A. where my family could be with me. It almost brought me to tears to change that M.O.S. number from flying to veterinary medicine, but it seemed the wisest thing to do. Once again the damper cooled the bird instinct within me and fueled the flame for caring for the animals. My return to the Basin would be delayed two more years but the dream was still alive.

In spite of the struggle and challenge of vet school I enjoyed my studies and all that I was learning. With the clinical aspects in the last two years, it took on a much more practical and useful meaning and I enjoyed it even more. We all worked to help each other from day one and by helping my study and clinic partners who were struggling to keep up, I seemed to benefit more than them. The repetition and review with them helped me, so when graduation time came I was about third in the class.

I don't think we realized how long and hard the struggle was until it was over. The relief was so great that the feelings of parting with friends didn't seem to evoke as much pain as former partings had done. Everyone just seemed to want to get away and pinch themselves so they knew they were not dreaming. We had to move our things back to Utah, and our folks, who had come for the graduation, helped us load and took us home. My stay was short-lived because I had to report to Air Defense Command at Colorado Springs. My home base was going to be Fort Ethan Allen at Burlington, Vermont, but I had to undergo eight to ten weeks of intensive training in administration, meat, dairy, and egg inspection. After a week in the office, I was sent up to Denver for two months of hands-on work in the procurement, inspection, testing, and quality checking of food for the armed services. We inspected the plants for sanitation, the processing for correctness and every aspect of food and its handling. I even became an egg candler for a few days and tasted enough butter and cheese to load a freight car. Joyce was able to be with me part of the time in Denver.

While in Air Defense Command headquarters, I learned—a bit to my surprise—how intense our armed services were preparing for a possible strike from Russia. My home base at Burlington, Vermont was a fighter interceptor base with several satellite radar bases under its command, which was all a part of a growing warning and intercepter shield all across our northern border and into Canada. We weren't playing games and the so called "cold war" was a lot hotter than the public was aware. I was to be the base veterinarian on a turn of the century calvary base that was being transformed into an air base and also do food and dairy plant inspection for the 1st Army area, which covered New York and all of New England. At the end of my training in Denver we loaded our little car and headed for Vermont with our two little girls, Rosalie and LaDonne.

My responsibilities as base veterinarian included the testing of the water once or twice a day, the inspection of all food that came to the base, and any food processing plant from which it came, if it was in our area. We were responsible for food sanitation, preparation and serving in the mess halls, cleanliness and spacing of bunks in the barracks, also with garbage and refuse disposal for all of the base. Rat, mice and mosquito control also came under our purview and I was allowed to care for the pets of Air Force personnel assigned there. It was a big job with a lot of interaction with many departments and personnel which required eight to ten copies of all the required inspection forms. I had three enlisted technicians assigned under me to help with getting the job done. I was once asked by Don McNiel on his breakfast show what veterinarians did in the Air Force. To explain it all would have taken too long, so I just said, "We take care of the horsepower in those planes."

It seemed like I just got everything organized and my men trained when I received orders to attend the meat and dairy hygiene school in Chicago. It was an eleven-week course during January, February, and March. Wintertime is not a good time to be in Chicago. It is a big place to keep clean and as winter ended it was very dirty and the weather was wet and foggy. One morning as we were looking out over the stockyards toward the city shrouded in haze and smog, Captain Buttons from western Colorado said in all seriousness, "If they were ever going to give the world an enema, this is where they'd stick in the tube."

Although the weather and circumstances left something to be desired, the school was the best in the world and gave me rare opportunities to see some of the world's largest meat packing plants, dairy processing plants, can factories, soup making plants, and egg and cheese processing facilities. Just to see the speed and skill of some off those meat cutters and the enormous volume of products being handled was a special privilege to me.

After the meat and dairy hygiene school, I returned to my base to enjoy the remainder of my tour of duty in what might be classified as country club style. Vermont was a very beautiful place and the need for a veterinarian in the little lakeside town of St. Albans was intriguing. However, our families were so far away that the beauty of Vermont never threatened our determination to go back to Utah. The real threat was staying in the Air Force. By the time my tour was up, I had ten years credit toward retirement and I only had to put in ten more. Col. Gorman, my commanding officer at Air Defense Command Headquarters, seemed to like me and twisted my arm pretty hard when he called on the phone. It was a tough decision because I could have retired as a Major or Lt. Col. at about the age of 41 and then gone into practice, but being non-drinkers we didn't fit into the social life and protocol in a comfortable way so we decided it was best to get out.

The original dream of coming to the Basin was partially shattered when I went into the service and Dr. Burritt, one of my classmates, decided to practice in Vernal. I watched with a sad heart as the first year passed, then heard that he had decided to leave and my hopes soared upward again. When the next class graduated and none of them went to the Basin, I was to the point of counting the days when the rug really slipped. Dr. Wesley Peterson, a graduate of the class ahead of mine, went to Roosevelt. That was the finishing touch. My dream withered and died, my anchor let go and I was floating. This was the time when staying in the service was a real temptation. Dr. Ben Allen, my clinic partner, called from Idaho Falls inviting me to come and help him out. His father had a heart attack and needed to slow down and he was alone in a busy practice. Another door opened to an opportunity that held great promise with someone I had come to love and respect. It wasn't hard for us to make this decision and start formulating our plans.

The Air Force would move all our things, so once again we packed our little green car. This time we had three girls instead of two. We thought it would be an opportunity to see places, so we went north into Canada. We plotted a course up the Ottawa River, across Eastern Canada and down to the ferry at Saulte Ste. Marie, crossing back into the U.S.A. again. From there we crossed Wisconsin, Minnesota, the Dakotas, Nebraska, Wyoming, Yellowstone Park, and into Idaho. It was an enchanting trip up the beautiful Ottawa River with the dense green forests and clear blue waters. Westward through forested uplands, we came to the well-groomed farms and small towns of central Canada. When the ferry docked and we drove onto the land of our allegiance it was good to be back in the U.S.A. The trip across Wisconsin was green with spring breaking out in every form. One morning we encountered swarms of Mayflies so thick they made the road wet and slippery as a half-inch layer of them accumulated. Minnesota with its forests, farms and lakes pointed us toward the great plains and left us filled with joy. The Dakotas, Nebraska, and Wyoming were the bread basket leading to the mountains and our anticipation grew with each passing mile. Our cup ran over as we we journeyed ever homeward and topped it off with part of Yellowstone Park and the Grand Tetons into Idaho, our new home. We couldn't have picked a more beautiful route and the kids traveled like troopers. When we got to Idaho Falls, the Allens had a house rented for us and our furniture had already arrived.

3

IDAHO INTERLUDE

The long trip had effectively closed the door on the Air Force and we were starting a new life once again. I was anxious to get involved and become acquainted and both Ben and his dad, Dr. J.S. Allen were just as anxious. They introduced me to everyone who came into the clinic and held me up as someone very special with great talents. I tried to live up to all their expectations and was especially blessed with the ability to follow a few penciled directions to some farm. Dr. J.S. and I hit it off well and I learned about client relations from him. Sometimes he would come and ride along with me on a call to show me the way, and all the time he would be telling me of experiences from which he had learned. He told me how exasperating some people can be but how, with great patience, they could get along.

One special experience was a nice warm day in late summer when we went out to treat a cow with a retained placenta. I was in up to my shoulders trying to loosen and remove the placenta and the odor was very bad. I had the owner holding the tail and as the cow strained, the fluid squirted out and the air behind it would make it spray. He was having trouble to keep from gagging and when I finished and was washing up I heard him say to Dr. Allen, "I don't know how you guys can stand to do that, it about gets to me, but I guess after awhile you get used to it." I'll never forget his answer.

"Well I don't know whether you do or not. I've only been doing it for 37 years."

When I first started working for the Allens they had what was left of a black Labrador retriever in their kennels. He was a real puzzle that we just couldn't figure out. He would only eat a taste or two of food and he had a very dark liquid stool that was limited in quantity. His temperature remained normal or below and he didn't act sick, except the weight loss and growing weakness had to be attributed to something. He was always happy to see us and would move from one stall to another and follow us around the back room. He'd been X-rayed, blood tested, cultured and examined over and over. He belonged to a man that lived in St. Anthony and after two months of hospitalization he stopped by one day and asked, "How is the dog?"

"He's still the same. Come on back, he'll be glad to see you." Dr. Ben led the way and we went back to the kennels. "I think he's getting a little weaker with time," Dr. Ben explained. The fellow dropped down on a knee and took the old dog's head in his arms. He still had enough strength to wag his tail as the man stroked and talked to him. His legs were weak and he soon sat down and seemed to plead for help. The man had a hard time swallowing the lump in his throat enough to speak.

"I think it's time to put him to sleep. I don't want him to

suffer," he said with his lip quivering and tears running down his cheeks. He dropped down and gave the brave old dog a final hug and we all went back through the hall to the office. He had regained his composure and we had all wiped away the tears. "There isn't any amount of money I wouldn't be willing to spend on him, it's just that I don't want him to suffer, if you know what I mean." He then explained that this was the only dog he had ever had or seen that would retrieve a duck in the water and take it to whoever shot it and not back to its owner.

Dr. Ben rehearsed all the effort and tests that had been run and told him how sorry he was.

"The only thing I can think of is cancer which doesn't show on the X-rays." The man was deep in thought for a time.

"If I don't call you by noon tomorrow, go ahead and euthanize him. See what it was, then call me and I'll come and get the body." We had all fallen in love with the old dog and found it hard to think about putting him away. The next morning I suggested to Ben that we prepare him for surgery and open up the abdominal cavity and see if there was something we could do. If it was the cancer we suspected and was too extensive, then we could put him out at that time.

When I opened him up I found a place where the gut had telescoped inside of itself for several inches and yet remained open enough to let a little fluid pass through in small amounts. The blood supply had not been compromised enough to cause necrosis and it was just essentially a stricture of the bowel. We cut out that piece and anastomosed the gut together and sutured him up. We were so elated and happy for the old dog that it was like a celebration. The owner called just after we finished and he was thrilled. We knew he wasn't out of the woods yet and we took every precaution to keep him on limited amounts of liquids while the gut healed. Almost from the first day the old dog was ready for food, and now instead of full pans sitting around, I think he felt we were starving him. Gradually we got him back on solid food and it was coming through him normally. By the end of the week he was eating all he could and it didn't seem to fill him up. It was so great to see him eat and begin to get his strength back. We all petted and loved him on every occasion and gave him free run of the clinic most of the time.

The owner came and took him home and we didn't see him for about six weeks or two months. One day the owner brought him back for us to see and he was all filled out with the shine back in his haircoat. I was in the large animal room when he came in and the old dog went through the clinic and finally came out to where I was. He trotted up to me and jumped up on my chest with his front paws and whined and talked to me like he was trying to tell me thanks. He seemed to know that we had helped him and happiness radiated from him as much as it did from his owner. That one incident seemed to make all the struggle to get through vet school worth it. I learned another great lesson: never give up.

The lesson of never giving up seemed to be epitomized with a call in the late afternoon of New Year's Day. Eastern Idaho was covered with a few inches of snow and it had drifted and blown, as it has a habit of doing, where wind comes as often and steady as it does in that area. The sun had already dipped below an ominous bank of clouds all across the western sky. It was chilly and I had overboots, coveralls, a coat and gloves but, I didn't have covering for my neck and ears except for the collar of the coat. As I traveled down the highway paralleling the railroad tracks, there were slick areas where the snow drifted across the road and I had to take it slower than usual. As I observed the cloud bank I thought to myself, "I'll be lucky to get back home before that storm hits." I wheeled up over the tracks and down a quarter mile lane to the dairy farm and pulled in by the milking barn. There, in a small pond of liquid manure and slush ice, was the cow. She was down with her hind feet out behind her like some that go down with back injuries. Her ears were lopped down and she could barely hold up her head. Out behind her hind legs was her uterus. It was completely inverted to a length of three and one-half feet and filled with enough intestines to fill a wash tub. I walked around the pond surveying the situation and determining in my mind that there was no chance of saving that cow. When I approached the barn, the owner came out having just finished cleaning the barn after milking. When he saw me there instead of one of the Allens, I could immediately detect a bit of disappointment or lack of confidence. It was the first completely prolapsed uterus I had ever seen and seeing the cow so close to dying gave me great apprehension, but the owner's reaction triggered an equally great emotion that said I had to prove myself.

"What are we going to do, Doctor?" he asked. I looked at a little high spot at the side of the pond where all the snow was blown off and thought that if we could get her up there, we might be able to do something.

"Can she get up at all?" I asked without much hope.

"Naw. she seems to be down in the back." I had a fleeting feeling that he wanted me to tell him to just kill her and get her out of her misery, but I wasn't sure.

"Have you got a tractor handy?"

"Yeah, it's just past the haystack."

"Well, the first thing we have to do is pull her up onto this knoll and then we'll see what we can do." He hurried out to get the tractor and when he came back I said, "Hook onto her head and pull her straight out here," pointing to the high spot. He waded out in that pond of manure at least a foot or more deep and hooked the chain. I was glad his boots were higher than mine because mine would have been submerged in there. When we got her out, her hind legs were still straight out behind and he held her head to keep her from rolling on her side. I leaned against her neck and suggested we get a bale of hay on each side of her which we did.

"Have you got warm water in the barn?"

"Yeah, there's plenty of warm water."

"We're probably going to need a lot of that. Have you got something we could put under the uterus while we wash it?"

He came back with a six foot square canvas. By this time the wind had picked up and a few snowflakes were starting to come down, but we stood on the edge of the tarp and pulled it up under the uterus. I went to the car for my quart cup in my calving kit and he came with two big buckets of warm water. I pulled off my coat and knelt down behind her. The sleet felt like needles poking me in the back of the neck and on my bare arms. I had him pour the cups of water over the uterus as I tried to rub and wash off the manure. The water ran down into a cold puddle on the tarp, then over the edge and down under my knees and legs, soaking through my clothes. By the time we had finished the first bucket of water the storm hit with a fury. A real blizzard pelted us unmercifully with hail and sleet. It was getting dark fast and I couldn't feel much difference in the cleanliness of that huge uterus. Manure was worked into the crypts on the cotyledons until I didn't think it could ever be removed. I thought "what's the use," but I asked him to flood it off while I lifted up on it. I knew the cow wasn't going to live if I didn't get it back in and maybe she would be able to slough out the dirt if I did. I was wet all over by this time and the wind was freezing my clothes, so I just lifted with my arms and pushed with my tummy. I felt the tension reduce as the intestines flowed back inside of her. I asked him to help me push what he could and I began to knead the neck of the uterus back inside. I'd push some in on one side then try on the other and when I'd pull out my hand it seemed to follow it out. I didn't think I was doing much good, but it got a little more pliable as I worked and I finally had the body inside and the horns started. By keeping one hand up inside, I could hold it in, while I put my other fist against the tip of the horn and pushed until there was pressure. By quickly pulling out my deep hand I could get a little more inside and slowly the thing began to go in. With victory so close, I forgot the pelting sleet and agonizing cold and shoved with all my might just as the cow mustered enough strength to try and push it out. She would push and I would just be able to hold it. When she relaxed for another breath, I'd gain a little until the uterus finally slid through the pelvis. I held it in with my left hand while I got my breath for a few minutes. Once again I felt the fury of the storm, so I began pushing and turning the horns back from their inversion. When I got that done, the cow relaxed and I thought she was dying. Her breathing was labored and her head had flopped down but she wasn't straining any more, so I hurried to the car to get a needle and some umbilical tape. She was still breathing when I came back, so I sutured her to prevent its coming out again. She didn't even flinch as I poked the needles through the skin. She was in deep shock and I expected her to die any minute, but there was a large pile of hay stems where he'd cleaned out the manger so I suggested we cover her up with that. When we had her well covered with a four- to five-foot layer of that hay, I went into the barn and washed off my arms and put my coat back on. My coveralls were solid ice by then and I was cold, but I did have one pint of electrolytes and some milk fever medicine in the car that I thought might help her. I burrowed my way down in the hay to her neck (we had rolled her out flat and pulled her hind feet up under her before we covered her) and I slowly ran the medicine into her jugular vein. By the time I had finished, she was holding her head up but that was all. I gave her some penicillin and covered her back up well and turned to the owner extending my hand.

"My name is Dr. Dan Dennis. I don't think she's got much of a chance in weather like this but who knows?"

"Thanks, Doctor. If she doesn't, she doesn't. I didn't think you'd even try when you saw her, but she's still alive."

"I'll be out tomorrow and if she's still around we'll give her some more penicillin. With all that muck in her uterus she's going to need a lot of help." By now my teeth were chattering and I was shivering, so I hurried to the car and headed for home. The blizzard lasted all night with snow and sleet piled up six or eight inches deep and the wind was still blowing the next afternoon when I bucked my way through some drifts to get into the place. The owner was in the barn getting ready to milk and the pile of hay was still over the cow.

"Is she dead?" I asked.

"I don't know, I haven't dug in to see, but I haven't seen any movement," he answered. I began to kick the snow and hay away from her back end and uncovered a little of her. When I put my hand on her she was warm and I could see her move as she breathed.

"She's still with us and feels warm. I'll get her another shot. Why don't you get her a drink of water." I gave her the shot and covered her back up again, then unburied her head which she had laid back around to her side. She took a sip of the water but didn't want much, so we just piled more hay over her and left her alone. I went back each day for a shot and the fourth day when I stabbed the needle in, she tried to get up. We pulled the hay back off from her and she got up and walked over toward the manger and the other cows like nothing had happened. Within a week she was coming back up on her milk and never had any more trouble. I think my challenge of measuring up to the Allens kept me trying when I otherwise might have given up. It taught me a great lesson that I have never forgotten. I know it's true in life—not just in treating animals.

There are times when you realize that what you can do will save the animals, but it isn't always the best solution. Hardware cases as we called them, were cows that had eaten some metal, usually wire or nails. When the cow eats, it merely wads up a mouth full of feed and swallows it without much chewing. When the metal enters the stomach it filters down through the contents to the bottom where it tangles in the lining and often penetrates causing pain and peritonitis. It can cause death if it penetrates toward the heart. It became a severe problem when the pick-up baler came into being. There was a man in Idaho who ran sheep on his place during the winter. His method of feeding was to drop the wire-tied bales and let the sheep eat them without cutting the wire. It seemed to prevent some waste from tromping. The result was a field full of bailing wire loops when the hay began to grow in the spring. He took an old dump rake and thought he had cleaned the field. He was tired of sheep and sold them to buy thirty-five dairy cows and built a milking barn.

We came into the picture when one of the cows got sick

and Dr. Ben diagnosed it as having a wire penetrating the stomach. At this time we operated on those cows to take out the wire, because the magnet was not yet in use. When we opened up this cow and cleaned out her stomachs, we ended up with at least a half cup of chopped up bailing wire. In a week or two another cow made the trip and she had a handful of wire too. The owner acknowledged that there was a lot of wire still in the field when he cut the hay. When the third cow was operated on with the same results he realized that all of them were probably loaded with wire because he had fed them all that hay. Operations were not the best answer for him. After that when one would show symptoms he would just take her to the sale and specify she go to slaughter. He started to buy replacements, but decided he'd better clean up the fields first. He sold the whole herd when fifteen or so had gone and decided to started over.

Hardware cases were certainly varied and sometimes amusing. A fellow brought a cow down from the St. Anthony area and as I operated, I reached down inside her and the first thing I brought out was a small metal fishing tackle box all crunched into a solid mass, but complete with fish hooks, spinners and sinkers still inside. I then pulled out three overshoe buckles, a belt buckle, and a large metal button like one from a uniform. Finally, I found a four-inch finishing nail that was penetrating and causing the symptoms. As I laid it on the bench by the other stuff, I said something to the effect that she ought to feel better after getting all that out and the owner kind of rubbed his chin and said, "The only thing I can figure is that the old gal must have eaten a fisherman." It was always good to have a sense of humor. It helped to accept the problems animals created.

Another day the phone rang and I picked it up, "Allen Animal Clinic may I help you?"

"Doc, I got a real sick cow. I think she's dying. Can you come?"

"Yeah, where do you live."

"It's five miles down the highway, then over the tracks just past those potato cellars, and a quarter mile south. I'm the yellow house on the left. Can you hurry? She seems to be having trouble breathing and grunts every breath."

"I'm on my way."

When I arrived I found a Jersey standing humped up a bit and acting like every breath was hurting her. She had a half moaning grunt on every breath.

"How long has she been sick?"

"Just now. She was fine when I milked her this morning and when I came out to feed right after lunch I found her. What do you think it is?"

"I'm not sure, but let me take her temperature and a few things. Maybe we can figure out what's the matter. Have you changed feed in any way in the last few days?"

"Nope, it's the same hay and that load of grain came better than a week ago."

"Have you seen anything in the corral like a bucket or something spilled?" By this time I'd found her temperature to be normal and her mucous membranes a normal pink.

"No, I haven't seen anything, but I can take another look," he said as he started to survey the corral again.

The old cow was a family pet and very gentle, but she was reluctant to move. I wanted to see if she were dizzy or showing any central nervous involvement, so I slapped her on the back leg and waved my other hand to make her move. She took a few steps and her grunting told me it hurt her. I pinched her backbone with my hands then pushed up under her left side with my knee and this really hurt her.

"She's really hurting ain't she Doc. What do you think it is? I didn't find anything in the corral and all the others are eating."

"Well, to tell you the truth I think she's got hardware, something poking through her stomach, but I've never seen one in so much pain. I think we ought to get her into the clinic and operate on her as soon as we can. Have you got a truck?"

"I haven't got a rack for my truck, but I think it would be easier to get her in that little trailer over there."

"Yeah, that'll work fine. You hook onto it and I'll get a rope for her halter." We backed the trailer right into the corral and the old cow seemed to sense that we wanted to help her and stepped right up into the trailer. I raced back to the clinic and had things mostly ready by the time he pulled in. We put her in the stocks where we operated and I shaved, washed and prepped the area. When I deadened her she didn't even raise a foot, the other pain must have overshadowed my needle pricks.

When I reached down into her stomach there was something about eight inches long and flat like a knife blade but the edges weren't sharp. It was wedged into the reticulum and one end was penetrating down toward the bottom. I pulled it out and it was a big spike that the kids had put on the railroad tracks and let the train run over it. It was curved a little on the end and resembled a miniature saber. I handed it to the owner.

"Well, I'll be. That thing was in the manger where the kids had been playing. I saw it when I dumped the grain in this morning, but I never dreamed a cow would pick it up, let alone swallow it. Boy, I'm glad you knew what the trouble was with her. She acts better already." It was a good feeling to know you had made the right diagnosis because so often you felt like you were on trial.

One day a man called me and said he needed help with a cow calving. There were three feet started and he couldn't figure out what to do. When I got there I washed off the cow and reached in following up the two legs that protruded the most. There were the hock joints so that made them back feet and then there was a head and the other foot. I felt back along it and determined that it had a knee so it was a front leg. It was, I thought, either really twisted and deformed or it was twins. By following back on the hind legs I could feel the stifle joints and hips and they seemed relatively straight so I concluded it must be twins. By pushing the one foot and head back into the uterus and pulling on the hind feet, I soon had one calf coming out backwards and with a little help it came right through. I then went in and straightened the head and found the two front feet. The second calf came out, followed by the placenta, which had come loose during the laboring process, killing both the calves. Well, it looked like my job was completed, so I gathered up my chains and lube and the man paid me and I was on my way. I hadn't any more than arrived and washed up when the phone ran. "Allen Animal Clinic, may I help you?"

"Are you the one that was just out here and pulled the two calves?"

"Yeah, that was me."

"Well you left one in her. As soon as you left she commenced pushing again and two more feet began to show. I pulled on 'em and out came another calf."

"Another calf! As big as the other two?"

"Yep, it was just as big as them."

"When that afterbirth came I thought she was all done. I didn't even feel inside of her. She had triplets, that's pretty rare in cows."

"Yep, she did. You goofed up on that one."

"I'm sure sorry I didn't check back in there. Is she alright now?"

"Yeah, she seems to be fine now, except she's worrying over those dead calves going from one to the other."

"Why don't you get the calves and the afterbirth out of the pen and I think she'll be OK." I thought, "She'll forget about those calves a lot quicker than I will, especially since he seemed to get so much pleasure out of telling me I goofed."

I guess everyone who is just starting out feels like people are kind of looking over their shoulder at them. I think I was a little more sensitive than most because I felt like I had to measure up to Dr. Allen and Ben. I did feel that I could perform the normal surgeries as well as they could and they must have felt the same because they had me do a lot of them. My first cesarean section was on the youngest heifer I've ever known to get pregnant. When I operated on her she was just thirteen months old and laboring on a full term calf. That meant she was only four months old when she got bred and that's very unusual even for a Jersey, which she was. The calf was alive and a good-sized one. When it stood beside her it looked half as big as she was. The owner was well pleased to have a live calf and save the heifer. He also enjoyed talking about this very rare occurrence.

When I finished up doing things for people I usually felt good about what I had done and sometimes there were comments that backed up that feeling. I remember one day when I was vaccinating sheep for soremouth. The sheep ranchers knew how soremouth could set a flock of fattening lambs back so they wanted them vaccinated before they turned them in the alfalfa fields in the fall. I had some experience behind me in this area while I was in school. We students were the free labor for the ambulatory clinic staff and it was our job to catch the lambs and set them on their butts and hold them while the clinicians vaccinated them. On this occasion I was to be the man with the knife and bottle of virus, and there were a couple of owners and half dozen herders there to handle the lambs. The procedure entailed making three shallow cuts or scratches close together on the inside of the hind leg where there wasn't any wool, then brushing some of the liquid vaccine into the fresh cuts. They would set the lamb on its butt on the ground, feet sticking out toward me. I would quickly make the little sliver with a scalpel blade held in my fingers. The object was to just cut through the outer layer of skin enough to bring a slight bit of blood. I would then take the little brush, which was submerged in the vaccine, and rub it into the cuts to infect this fresh wound and cause a lesion that would scab and produce immunity.

It was about ten o'clock in the morning when they finished separating and counting the lambs and we started vaccinating. The group of them could keep a line of lambs set up and waiting, so I moved from one to the next while the finished lamb was then taken to the gate and released. There was a huge pen full of lambs and they thought it might take two days to vaccinate them, but I worked very rapidly and kept them busy. We didn't stop for any lunch except some punch and cookies the women brought and by the time the sun was low in the west we had finished all those lambs. Needless to say, my back was tired and cramping after bending that many times, but so were the other's from handling the lambs. I straightened up and put my hand on my hip and bent back as far as I could. It sure felt good and the owner walked over and shook my hand.

"You're a good man, Doc. I never dreamed we'd get 'em all done today, and I've never seen a better job done either. Do you know how many we did?"

"Well, I don't know the numbers but I feel like it was enough. Do you have a count on them?"

"Our count was 2,223, plus or minus one or two."

"No wonder I thought it was enough. How about if I just round it off at 2,220."

"Sounds fair enough to me. Let me make you a check. We've never been able to do all of them in one day before, but you were super and I sure thank you."

I learned to appreciate the livestock owners. They were good down-to-earth types and I always trusted them or felt I could at least. At the same time, I learned that you can't always trust an animal. One day a man pulled up behind the clinic with an old Jersey cow in a two-ton truck. He had brought her in from Mud Lake and said she had dropped off on her milk production all of a sudden. I thought I could climb up there in the truck and examine her so I said, "Is she gentle or will I need a rope?"

"She's gentle as a dog. The kids just go out in the field and milk her wherever she is. She even lets 'em ride her."

I put my thermometer and my stethoscope in my pockets and climbed up over the side of the truck and started backing down the inside. About the time my feet were approaching the floor that old cow hit me in the butt and threw me up in the air. I hung on to the top of the truck and tried to get a toe-hold on the side and she smashed into my legs and feet. I finally got away by pulling myself up and flopping across the truck rack on my belly. Boy, was she ever upset over something.

Another valuable lesson I learned in Idaho was the value of the Kingman tube for treating bloat. The Kingman tube was a large stiff rubber hose about one and a half inches in diameter containing some webbing and was about eight feet long. It was named after Dr. Kingman from the famous Wyoming Hereford Ranch where he was herd veterinarian. In the late fall and early winter, cattle were turned into the sugar beet fields to clean up the beet tops. They were very good feed, but not infrequently the cattle wouldn't chew the beet top sufficiently, and as they attempted to swallow it the piece would lodge at the entrance to the stomach or in the neck at the chest opening. This obstructed the passage of gas up the esophagus and caused the animal to bloat and sometimes inflicted considerable pain. The pain symptom was valuable because it frequently alerted the owner, making it possible to get help before the bloat got bad. The Kingman tube could be used as a rather safe ramrod to push the beet top into the stomach without rupturing the esophagus. It was also very valuable in relieving other types of bloat as well.

I was well accepted by the clientele of the Allen Animal Clinic in Idaho Falls and both Ben and his father were pleased with my performance. They were talking about a partnership arrangement and we were looking at homes. My dream had flared and dimmed as the years went by, but now it was time to make some permanent decisions. The opportunity with the Allens was exceptional and with someone already practicing in Roosevelt the circumstances were rapidly cementing my future to eastern Idaho. I had about resolved to stay there when word came that Dr. Peterson was moving to American Fork. The old fire seemed to flare up again and the difficulty of decisions plagued me again.

4
STARTING ON MY OWN

New Vet In Town

I hated to leave a sure thing for what seemed a questionable future in view of seeing two qualified veterinarians come to the Basin then leave within a year. Maybe there wasn't enough business there to make a living. I was really torn about what to do. I had property which my dad was farming and he seemed to be having health problems. It was a tough decision but when I went to check things out, especially with my father, I discovered he had symptoms of diabetes. He said he'd been to a doctor, but I convinced him to go in again and told him to make sure they checked his urine. Sure enough, they found it was diabetes and they put him in the hospital to get him regulated.

I guess the closeness to my dad outweighed my love and loyalty to the Allens, so the scales were tipped once again and we decided to make the move. I arranged for the house Dr. Peterson had used for his home and office and prepared to move in April of 1954.

I talked to the Allens about buying some basic instruments and a few drugs to get me started and they were most helpful and gave me some good counsel and advice as well. When I wanted to settle up with them they wouldn't take a cent from me, we just shared the hugs and tears. Dr. J.S. Allen had a friend in town that made him several sets of dog kennels and he said he'd have him make me a five cage set if I wanted. I didn't have any available in Roosevelt so I took advantage of the offer, and it was one of the wisest decisions I ever made.

After vet school and my experience in Idaho, I wondered about practicing in the Basin. I knew that of necessity the people there had been steeped in resourcefulness and were used to doing whatever was needed by themselves. They were isolated in an area where making a living was difficult and they met the challenges head on. If an individual couldn't handle a problem he called in a neighbor, and if he couldn't do it, then it was left to the community. I knew how self reliant they were because it was a part of me also, but I knew that I could help them if they would let me. I knew how frugal and close they were with their money because they had to be in order to survive, and this was a concern of mine. I knew I had to prove to them that my services would save them money in the long run by reducing losses. That part was challenging, but I also knew how willing they were to help, how honest and trustworthy they were, and how the day was never too long. I looked forward to becoming one of them, of learning from them as I went along and being able to share my acquired knowledge with them. To me they were the salt of the earth.

We came to Roosevelt with the dream burning brightly and brought with us a treasure trove of practical knowledge and ideas that the Allens used. The stage was set for the beginning of my own practice. We put an announcement in the local paper and waited for the phone to ring. For much too long the days were mostly silent or punctuated with a few inquiries on what it cost for this or that. I couldn't as much as sell a spay on a cat for ten dollars. The usual comment was, "I'll shoot the thing before I'll pay that much." It was more of a blessing than we'd ever realized to have milk, eggs, butter, potatoes, and canned fruit and vegetables from the farm. We struggled to pay the rent, phone bill, and buy gasoline, but we had faith that the need for our services was out there if the people only realized it. Was the dream just a dream after all that would never come true? This thought ran through my head over and over. I think the embarrassment of having to ask the Allens to take us back and the need to help Dad on the farm kept us trying.

I contacted the sales barn and they were anxious to have my services so they could be accredited to receive out-of-state cattle and other privileges accorded. That work provided us with fifty dollasrs per week. I felt it would be a good way to get acquainted with the livestock people—which it was—but by having to enforce the vaccinating and testing regulations I was often in an adversarial role and the contact didn't always end up in a positive acquaintance. I tried to maintain a good sense of humor, to be sympathetic to their point of view and to never exercise my authority in a dogmatic manner. As the years passed I was able to earn their respect.

Small animal patients finally did begin to trickle in; I had a few sick dogs and spays. There isn't anything more pathetic than the terminal stages of a dog with distemper. Few would even pay the few dollars for a painless euthanasia because a bullet was cheaper. Spays were mostly the result of a female dog in heat with about twenty-five males hanging around, especially if the kids had fallen in love with her. The people were more than pleased with the recovery and solution to their problem.

At first the large animal calls were mostly telephone inquiries about animals that had died or were sick, but they only wanted my advice. No one ever asked me how much they owed or offered to pay, but I didn't mind because my desire was to be as helpful as I could. The thing that hurt a little and busted my ego balloon was their listening attentively to what I had to tell them and saying. "I don't think that's what it is because Dave Huish said he thought it was this." They had relied on the local druggist for so many years that they took his advice instead of mine. At times I used to wonder why I spent those eight long years going to college and how nice it would have been to be back up in Idaho where I had a measure of respect.

Sometimes I could convince them to let me come out and do a necropsy to determine, if I could, why the animal had died. I guess they figured I couldn't hurt it if it were dead. I only charged the mileage charge on my car and did the cutting up of the dead carcass for nothing. When I was able to actually show them what killed the critter, it made believers out of them. My credibility gradually increased and my pathology skills and knowledge improved as well.

I had a very good relationship with the local by-products man. He picked up the dead animals and I examined them as he cut them up in his plant. It saved the ranchers money, got rid of the carcasses, helped the by-products business and I could often diagnose the cause of death. Whenever I was called out to necropsy an animal I would suggest they call the by-products man to get rid of the diseased carcass, and when he picked up dead animals he often suggested that the owners call me to go look at it if they didn't know the cause of death. He was a great help to me and I did all I could to get people to use him.

At first it seemed like all the sick ones I was called out to see were about ready to draw their last breath. They had been given every medicine and remedy that the drug and feed stores sold and home remedies from all the neighbors in a five mile radius. When I couldn't save the animal, the owner felt the vet just cost money. Occasionally, when visiting one of the hopeless critters at the end of the trail, I'd see something in another one that I could help them with and gradually I felt my standing moving up a notch. Thank heaven for milk fever cases where I could perform a miracle if the owner hadn't caused a fatal case of mastitis from pumping the udder full of air. Milk fever was a misnomer because the problem was a hypocalcemia, or low blood calcium level, which routinely produced lower than normal temperatures. This lowered blood calcium level also effectively paralyzed the large body muscles and the cow couldn't rise. It was caused by the udder taking calcium out of the blood to produce milk faster than the body could mobilize it from the bone. During the dry period the body mode was to store calcium and other minerals in the bone. Then, with the sudden shift in the hormones at birth, the udder shifts into production, pulling heavily on the calcium in the blood. It takes a day or so for the mode to shift from storing to mobilization from the bone. The result is muscles getting weak and paralyzed and the cow going down, unable to get up. It could progress to the point where respiratory muscles failed and the cow could die. My role was to administer a solution of calcium into the bloodstream in sufficient quantity that the muscles would again be able to contract and function. The cow oft-times would recover and get up on her feet while I was treating them. It was always my caution not to milk the cow for a day to allow the body to catch up. Well, that's a simplified explanation of a very complex metabolic problem, but I was all right in their eyes when a very depressed, paralyzed cow awakened and stood on its feet.

Apprehensions And Concerns

Having come from Idaho where vet service had been available for many years, I sometimes got very discouraged because the public was so unaware of the value of professional veterinary medicine and how important it was to them. I couldn't blame them, because they had lived without the service in the area and had no way of knowing. I think some of them felt I was just someone else making a living off from them in a parasitic way, so they avoided me as much as possible. Others felt I was a last resort when all else had failed. This was very frustrating because it was usually too late to save the animal, but I was able to make a diagnosis sometimes and advise them in the area of prevention.

I think there was a widespread feeling that medicine was a lot cheaper from the feedstore, drugstore, and farm-to-farm peddler, than getting it from me. However, this was not always true when quality and care in handling were considered. I tried to stay competitive on the common things used, and sell the higher quality products as cheaply as I could, but very few people ever bought from me. Sometimes they would come in to get my advice for which no charge was ever made, then go somewhere else to buy the medicine. Those few pennies they might have saved almost made us look elsewhere when we were struggling. Certainly a little monetary help would have made it possible to purchase the equipment necessary to do a better job. It took ten years to make enough to buy a used X-ray machine for under $500.00, and over fifteen years to get an autoclave. Laboratory diagnostic equipment was not to be mine until after my son came back to help me in the business thirty years after I started.

Some clients did support me, and as the years passed more and more people learned the value of good advice and the proper medicine to go with it—and I certainly appreciated them. It wasn't necessarily the small margin of profit that counted as much as the vote of confidence it gave me. When the costs of inventory investment, refrigeration, losses from aging, and dispensing costs were counted, you don't make a lot from selling pharmacuticals, but it's very satisfying to be able to send a client out the door with what you think is best for them.

One other facet that aroused some bitter feelings was the assumption by the public that my income was high. If I had taught school, I would have made more actual take-home pay than we had. It was necessary to purchase and furnish all of my instruments, equipment, vehicles, facilities, and everything I used over the years. My only retirement income came as the result of real estate appreciation and being able to sell some property, and not from the profits of my practice. I knew some small animal and equine practitioners made good incomes, but those of us who depended primarily on food animal practice had to watch our pennies. I tried a few enterprises on the side to augment what I made, but the time constraint of being on call twenty-four hours all the time limited what I could do. All I could do was wish I were rich like people thought I was.

Smelly Encounter

I had gained some experience in delivering calves in Idaho, but the requests for my help in the Basin were slow in coming. I was a pioneer among people who were used to doing for themselves. They had never had professional help to call upon and just didn't know what could be done. I also found that it was difficult for many livestockmen to swallow their pride and admit that I might have some knowledge and skills beyond their own. I think some of the first calls I received were from owners with a rotting, dead calf to pull and a stomach so weak they just couldn't handle it. At first many of these animals were shot to solve the problem. Once in a while, after the owners and all the neighbors couldn't get the job done, I'd be given a chance. With all the traumatization and effort that had already gone on, my success wasn't the best many times either. The word that I could help didn't get around very fast and it was a bit discouraging. Time taught me that client relations were a two-way street and the management and correction of calving problems was an evolutionary process. The owners had to learn to call on me early enough that lives could be saved and I had to learn to take each case as it came and do the best I could. I battled two erroneous schools of thought, the first was to hook onto the calf as soon as possible and pull, the other was wait and see until it was too late. Then there were those cases where a cow slipped away and wasn't found for a day or so.

One case I remember well was at Cecil (Doc) Jenkins' place. Doc was a special friend from the days we attended high school together. He had a herd of cows and was a breeder of fine quarter horses by the time I came back to the Basin to practice after fourteen years of war and school. He was a great rodeo man and did a lot of calf roping. Everyone in the country knew him and as far as I could see they all considered him their friend. He was a leader in high school and when I came back he was still the same—jolly, fun loving and not above a prank now and then. I did wonder how he felt about me. Could he accept me and not wonder how that big dumb kid from Myton could be a veterinarian? Whatever his feelings were, we were still friends and I appreciated that.

One afternoon late in the day I received a phone call. "This is Doctor Jenkins." I don't know where he acquired the title of Doc, but he always emphasized it when he called me. "I've got a cow that can't calve. Could you come down and help me?"

"I sure can. Where is she?" He hadn't offered any explanation about the cow, but it was only a mile away, so I didn't ask any particulars. When I arrived there was the cow down in the back of his truck. We grabbed her tail and back feet and slid her out to the end of the tailgate where I could work on her and she didn't struggle or object. I learned in the process that the calf was dead and well putrefied. When you work on one like that, you don't appreciate being swiped in the face by a slimy tail, so I handed it to Doc. When I reached inside, the calf was bloated up and I knew it couldn't be pulled normally. I took my hook bladed knife out of my kit and pushed my hand in as far as my body would let it go and began to slice the skin on the dead calf from the point of the shoulder blade down to the foot. It was a still afternoon with no breeze and no way to escape the ripe odors. As I worked to loosen the skin on the leg going in and out I noticed Doc turning one way and then the other to look the other way. I couldn't blame him. That odor was plenty strong. When I finished cutting the skin around the leg above the hoof, I put a chain on it and pulled it out with my puller. Now we had a dead leg out on the ground and the smell of the cow as well. I thought Doc looked a little "green around the gills," as they say. He had a neighbor, Ned Gines, with him. Ned had gone with Doc to help load and bring the cow down when Doc had found her hidden in the willows. Doc kept moving farther and farther back as we worked and was out to the end of the tail at arm's length. They kept a conversation going like two excited women. I could tell they were doing their level best to keep their minds involved in anything but the dead calf. I chained up the head and the remaining foot, applied a copious amount of lubricant around the calf and began to pull. The calf came out to the hips then came apart and flopped on the ground.

"Ned, you better take this tail, I can't take any more," Doc said as he bolted away about ten yards and began to take a few deep breaths.

"I don't think I can take it either. Dr. Dan, how in the world can you stand to work on her? Doesn't it make you sick?"

"It doesn't make me sick, but I can't say it doesn't bother me. It's just one of those jobs you have to do." Ned was now holding the tail at arm's length and I was back inside trying to turn what was left of the calf around and get hold of a leg.

"You must be made of cast iron, or have a cast iron stomach at least," Ned offered through half clenched teeth.

"One thing I can tell you is that when you get all through and can't wash the smell off, it teaches you to eat your bread with a fork," I chuckled. I guess the thought of eating tipped the scales because Ned began to wretch and throw up and as soon as he did Doc followed suit. They were gagging and having the dry heaves all the time I was pulling the rest of the calf. I think Doc gained a measure of appreciation for me that day.

First Facilities

As the practice grew, some of the bovine clients began to haul their animals in to me, and I needed a place to unload, examine, and treat them. I went to the farm and brought some cedar posts which I planted deep in the backyard. With a few 2 x 6s I made some stocks with a head catch on one end and a small corral with a loading ramp on the other. It was a rather meager beginning with only enough room for about two cows but it was all I had room for. It was located right there on main street, and I wasn't too sure how the public and the city fathers were going to look upon it. I only used it to unload and treat an animal, then load it up again, but I worried about getting a wild one that might escape. One thing that was in my favor was that the well-worn old trucks with racks wired together probably couldn't have kept a wild one in it long enough to get to town. We were only in that location for a few months, but sometimes an operation on a cow would attract an audience.

I remember doing a hardware operation there that caused the owner great elation at first, with sadness that came later, and was a great lesson for me. It was a Jersey family cow and, like most, a special pet to all the family members. I clipped the hair from the surgical site, scrubbed it and disinfected the area, then began to deaden her. I used a paralumbar block, and some of those in attendance cringed and turned away when I stuck the long needles down into the back to deaden the nerves. This was a good way to anesthetize a critter because once it was done they never flinched or moved as you cut down through the skin and muscles into the stomach. It caused the spine to bend a little as the muscles on one side relaxed helping to open the incision area. I had a metal ring of brass welding rod about a foot in diameter with hooks welded around the inside which held a rubber ring. On the rubber ring I had a series of hooks that I could pull the stomach up to and attach the hooks which would hold it open so I could reach in and come out without fear of contamination. It also made it possible for me to do the operation by myself. When I reached down to the bottom of this cow's stomachs there was only one smooth piece of bailing wire about four inches long, but it was moving with every beat of the heart. It had penetrated forward through the wall of the stomach, the diaphragm, and into the heart. I pulled it out and showed it to them and explained that I'd removed it from sticking in the heart. They were elated and the kids even cheered. I explained from what they taught us in school, that once the pericardium was penetrated the cow would likely die and suggested that she could be slaughtered in a few days. In their eyes there was no way that was going to happen, they were sure she would get better. There was very little adhesion of the stomach wall and the wire was smooth and new. It was my first actual experience with a heart puncture, and I even thought she might live. The bend in the back made the incision a little harder to close, but it all pulled together and she looked as good as new when I finished. They were all sure she was going to be all right, and it certainly appeared that way the next week when I went out to take the stitches out. She was eating good, back up on her milk and seemed as normal as ever. At the end of another month however, she was beginning to swell in the brisket and I knew the end was coming. It was a sad lesson for us all, but they said they wouldn't have done it any other way.

The Ideal Truck Bed

When we first moved to Roosevelt we lived next door to Crumbo Motor. They were very handy so I took my car there for oil changes and grease jobs. I got to know Albert Crumbo and Bill Murdock very well and I was very pleased with their work. Whenever I needed something, they would move me ahead of everyone else and get me in and out in as short a time as possible because they knew the nature of my work. One day as I pulled up at the house, Bill waved me to come over. He said he had something to show me. When I got inside he took me out in the back parking area and showed me an old Ford truck with a telephone company bed on it. There were doors on the sides with shelves, divided shallow drawers that pulled out, deep drawers, and a cover that could be pulled down over the open back part. It was ideal for carrying all my medicines, instruments, calf puller, ropes and there was room enough to haul a calf if I needed. They said they had just taken it in on a trade and could sell it to me for about $495.00. The truck had over 100,000 miles on it and the U-joints were worn and complaining but it started and ran well. I didn't have the money, but I found out that having that "DVM" after my name surely mislead the bankers to think that I soon would! All I had to do was sign my name, and I had a practice truck I wouldn't have traded for any available at that time. It was super and I transferred that body onto quite a number of new trucks as the years passed. It was distinctive and people knew me from that time on. I built a 20 gallon stainless steel water tank into it and that really gave me everything I needed. I knew that Albert and Bill could see how hard my practice was on my car, and that's why they thought about the truck for me. It was a blessing of immense value.

Many times when I was called out to see and treat an animal, it would be out in a big pasture with no corrals around. I would get the owner to drive my truck and I would stand up in the back and rope the patient. We would then snub it up to the truck where I could work on it. I usually had a number of dents and once in awhile I had a broken headlight where the animal's head had hit when it came to the end of the rope. The owners used to tell me I ought to have a saddlehorn on top to dally my rope to. There were other times when I could get close enough on foot to throw a rope on the critter, but I couldn't hold it. I'd get in the truck and run a front wheel onto the trailing rope and stop. This would hold it until I could get another rope fastened or the nose tongs in its nose. There were times when this was a little dangerous. One time I roped a horse that was supposed to be halter broke, but I couldn't get him to stop when he came to the end of the rope. I followed him down the lane in the truck trying to get close enough to drive onto the end of rope. He was trotting along at a pretty good clip when I hit the rope. It immediately stopped and flipped over in a somersault. I hit the brakes hard and almost crashed into him before I could stop. I didn't end up on the rope, but he decided I meant business anyway and stood waiting. He let us lead him anyplace after that sudden stop.

The back of the truck was often an escape route or a place where I could drop a rope on a belligerent critter that wanted to get at me. Sometimes I wondered if it was going to climb in there with me. Being that close made it easy to catch; however, getting the rope off sometimes posed a problem.

Horse Castrations

It was a little tough getting any requests for castrations. There were two untrained men practicing veterinary medicine in the area without a license and more experts at castrating horses than there were communities in the Basin. There were a lot of horses but I didn't get to do many castrations. Occasionally, I'd be called when one would swell up or a leg was broken because they hadn't tied them properly.

From my own castrations and the complications I observed with others' efforts, I determined that most of the problems were the result of inadequate drainage following the surgery. I decided that I would remove a sizeable portion of the entire scrotum instead of making two separate incisions over each testicle. When I used that method it did work out better and it became my routine. Whenever one of the local neighborhood experts had occasion to watch me, they observed in horror, for they harbored superstitions that the scrotal septum should not be violated. I'll never forget one of the older more emphatic ones declaring, "If that horse does live, he'll never be worth a damn because you cut the septum between the testicles."

Last Struggle

My efforts didn't always end up like I wanted. I remember when a good friend, Garth Anderson, called me down to castrate a two-year-old stallion he had raised and one in whom he had great hope because of its spirit. This was before the days of short-acting chemical restraint so we just put a rope harness on the animals and pulled them down, tied up the legs and did the castration. This was the procedure we used on this horse. Although he was strong and gave us a tussle in throwing and tying him, everything went as planned and I got his legs doubled and tied so he couldn't injure them. I did the surgery and clamped off the cords with my emasculators and there was a minimum of hemorrhage. Just as I began to untie his legs and the other men let him roll back over onto his side, he went into a terrible struggle to try and free himself. As he finished struggling he seemed to relax so I quickly removed the ropes. He didn't make any move to get up and in a second or two went through a couple of death gasps. His legs kicked once and he was dead. We were all shocked and they turned to me for an answer. I didn't know, but I suggested we open him up and see if we could tell what had happened. I suspected it was just a heart failure and wondered if we would see any evidence that would confirm why it had happened. When I opened the abdominal cavity things seemed a little pale but otherwise normal. We then got Garth's axe and I began to chop off the ribs and open the chest cavity. It was all full of blood. When we pulled the lungs and heart up we found the aorta split about six inches on the outside of the arch as it came up out of the heart. It may have been weakened by a developing aneurism or something but he had just struggled so hard that he burst it in the process. I was shattered to think we were all finished except taking off the ropes. Garth said that it was typical of his spirit. I knew that it was a very rare occurrence, but it caused me a lot of concern.

Sucostrin

The general anesthetics we had were not very practical and took a long time to wear off. The horses fighting to get to their feet would keel over a few times and it was a problem. It was shortly after this when sucostrin became available. It was a curare-like medicine that paralyzed all the muscles for a short time. It had been used in human medicine some and veterinary medicine picked up on it. The dosage was critical and it only took three cubic centimeters or less to knock an average-sized horse down. It only lasted about one or two minutes, but for me it was ideal for castrating horses. You had to inject it in the vein and within five to ten seconds the horse was down. I usually had my emasculators in my pocket and the scalpel held in my teeth. The owner or helper could easily pull the top leg forward because all muscles were relaxed, and removing the testicles was equally easy because there was no pull on them. When I was through, they dropped the leg and I'd check to see if he was breathing sufficiently. Because the respiratory muscles were paralyzed they sometimes didn't start breathing normally. If that happened, I would throw my weight onto my knee which I placed just behind the horse's ribs. This would cause an expulsion of air from the lungs followed by a reflex inhalation, after which breathing would begin again. After a few breaths to overcome the anoxia, the horse got straight up and usually went to eating grass if any was available. In about three minutes it was all done and the horse had no after-effects from the medicine.

It was comical to do unbroken, wilder horses. We would run them into a chute and I would begin to talk to them and pet them on the neck. Once I could touch them enough to hold off the jugular vein (which surprisingly didn't take too long), I would flick in the small needle and most didn't act like they even felt it. I would then connect the small syringe with about two-and-three-fourths cc's of medicine and squeeze it into the vein. I would holler okay and they would swing open the gate and out would go the horse. He would make about two or three jumps toward supposed freedom and go crashing down and we would already be on our way to get to him.

One day Dee Allred had six or eight stallions in his bucking string to castrate. We put them in the rodeo pens where the riders got on. I stood up on the catwalk on the inside and Dee was on the gate. When I'd give Dee the signal he'd swing the gate and the horse would start out into the arena. I'd vault the fence and follow him out and when that horse would go down Dee would just burst out in laughter. It was really funny to watch how much of a kick he got out of it.

As with every innovation, I learned some things about the use of this medicine. First, by giving a little less than the recommended dose the horse had a quicker and safer recovery. Next, you couldn't give it to a horse that was pulling back on a rope choking itself.

I made a bellows with a rubber tube I could stick up one nostril, and by holding the other nostril closed I could force air into their lungs, effectively giving them artificial respiration. On those that were compromised or didn't respond normally, it was a very effective safety measure, but there was still some risk involved.

The main problem with the drug was that it didn't have any true anesthetic property and although the horse couldn't move, it must have felt some pain. No one felt good about that aspect and as soon as other short-acting anesthetics came out we stopped using it.

With the critical dosing levels and no warning or corrective action possible on those animals with existing heart or lung problems, I was always on the edge of apprehension. One day Steve Esauk, a Utahn, called with an older stallion to castrate. It belonged to some friend out on the Wasatch Front and Steve was keeping him on his small ranch. I arrived and looked him over. He was a good-sized black Thoroughbred with mannerisms showing plenty of spirit. I calculated that he would need a little heavier dose than most and reasoned that because of his age he might have a tendency to bleed. I went ahead as with all the others, clamping the cords as long as I dared. When I finished, after taking a little longer than usual, he was still not getting deep enough breaths. I pushed him in the side and got him started and it wasn't long until he stood up. I was washing my hands and equipment and Steve looked over at the old horse with a smile and said, "I'm sure glad that's over. You know he wouldn't take $10,000 for him." I thought to myself, "Now you tell me, after it's all over." I was thinking he was just the $50 to $100 horse like all the rest. It was probably a good thing I didn't know.

Sucostrin was good medicine in a way, but very strange.

One day the highway patrol called me to come see a horse that was hit by a car over at Half-Way Hollow between Roosevelt and Vernal. When I got there I found the horse down with a broken back and everyone gone. I didn't have a gun or euthanasia medicine but I had a couple of bottles of sucostrin. I took five times the usual dose and figured it would kill the horse for sure. I gave it all in the vein and he stopped breathing as usual and I thought he would die, but after awhile there were tiny shallow respiratory motions and he was coming back. I hurried and gave the second bottle and waited. It wouldn't kill that horse, so I had to euthanize him a different way. On a number of other occasions I tried to euthanize horses with sucostrin and never could, but it killed a few when I didn't want it to.

Another interesting thing I learned was that wild horses captured from the desert, the original wild mustangs, had a tolerance of twice what other horses had. It took two and sometimes two-and-a-half times as much sucostrin to make them go down.

Grant Pickup had found an orphaned colt out on White River and brought him home to raise. It didn't grow to be very big, but in the process his daughter became very attached to it and wanted it for her own. It was a stallion and as it matured it naturally became a pest with the mares and other horses. One day he called me to come down and castrate him. When I walked into the corral and checked him he was about the size of a big shetland and I mentally calculated about how much medicine he would need and proceeded to get on with it. With the emasculators in my pocket and the scalpel in my mouth, I gave him what I thought he would need. The membranes rolled back in his eyes and he staggered a little but he never came close to going down. I waited for the effects of the medicine to wear off and pulled up enough medicine for an average sized horse. When it hit him, I tried to pull him sideways and spin him around and did everything I could to get him to fall, but he stayed on his feet. I was puzzled and didn't want to overdose him so I waited quite awhile to be sure the medicine was all out of his system. I then gave him nearly two times the dosage for a thousand pound horse. This time we were able to get him down, but even then he struggled some as I did the surgery.

Hidden Seed

The cryptorchid or hidden testicle horse, where one testicle remains inside of the abdominal cavity and doesn't descend into the scrotum, always presented problems. The undescended testicle in the warmer environment inside the body is infertile but it produces more than a normal amount of hormones. This results in a horse that has higher levels of libido or sexual activity. They are meaner and harder to control and more treacherous than a normal stallion. The owners always referred to them as being "more studdie" than a normal male. I was usually presented with one where the local expert had removed the one testicle but couldn't find the other one. It was my job to go up through the abdominal wall at the inguinal ring, and find and remove the offending source of trouble. The risk was greatly increased because of the ease with which the intestines could herniate out through the hole you made, and my first attempt proved this.

I got a call from an owner over in the west side of the Ashley Valley and stopped by to see the horse. He was a big black horse weighing about twelve hundred pounds. He was gentle and well broke but they couldn't handle him around other horses. I explained the risks and how we would need additional help to do him. They were quick to explain that he was no good like he was and if we lost him it was all right; they wanted to try. We set a time for a few days later when they could get help lined up and I left. On the appointed day I had time enough on the way to the ranch for a few thoughts of trepidation but knew I had to do the job no matter what the outcome.

We put on the throwing ropes and I gave him enough anesthetic in the jugular vein to make him go down. I doubled and tied his legs and then we rolled him onto his back. My first challenge was to determine which testicle was retained. By close examination of the scrotum, I was satisfied as to which side had a small almost imperceptible scar indicating where the other testicle had been removed. Once I was sure enough to proceed, I washed and disinfected the area. I made an incision a few inches anterior to the inguinal ring as we had been taught in school, and separated the soft tissue with my fingers. Once I had an opening in the ring to admit a finger, I searched and searched for the gubernaculum testis, a small cord-like structure that is supposed to guide the testicle into the scrotum. We were told if we could pick it up and pull on it the testicle would come to the opening. I couldn't find anything and after putting two and even three fingers inside, I still came up empty handed. I had to go all the way, so I forced my hand through and began to search. All I could feel was the multitudinous volume of intestines of every size. Sometimes I caught hold of something that felt like it could be a soft flaccid testicle and I would pull it out only to find it was intestine. The time went on as I probed here and there and ever deeper. I was quiet at first but I finally confessed that I was having trouble and the men comforted me with words of encouragement. Finally, with my arm in almost past the elbow, I felt something different down next to the kidney along the back. I grasped it in my hand and pulled it out, and sure enough it was the culprit I'd been searching for. I clamped it off with the emasculators and thought, "Now, if I can just keep the horse alive." The hole was big and the intestines floated up into it freely. I remembered in school they had told us to put a gauze pack soaked in iodine in the opening and hold it in place with a couple of skin sutures. We were to wait a few days until there was swelling and pull it out. I got a big role of gauze and had them pour iodine on it as I wadded it into my hand and pushed it into the hole. When it was filled, I placed the end out through the skin so I could grasp onto it and pull out the pack, then I sutured the skin in two places. We let the horse get up and everything looked fine. I'd passed my test, even though it was a difficult one, I felt good about the outcome.

Four days after the operation I felt it was time to remove the pack. When I arrived there seemed to be quite a lot of swelling, so I petted the old horse and rubbed his flank and abdomen until he let me get my hand back to the incision. I found the end of the gauze and gently pulled and it snaked out through the hole. The old horse was very patient and didn't move at all. When the end of the gauze fell away, to my horror the small intestines began streaming out until there was nearly half a tub full. This caused discomfort to the horse and he began to move around. I didn't have any hope that I could save him but I was compelled to try. I ran to the truck and got some anesthetic while the owner tried to control the horse. By the time he was down and asleep enough to work on, those guts were covered with manure and dirt. I sure wished then that I had taken him out onto the grass where we'd operated, but not expecting any trouble I went ahead there in his pen.

Things were so dirty that I didn't feel he had a hope of surviving, but once again I had to try. The owner got a wash tub and we filled it with water from a tap and some jugs of warm water that I had in my truck. It was cool and I thought the shock would be great, but we tied his leg back and got the tub under the filthy mess. I washed and washed and we got clean water and washed some more. At length I began rinsing a section and pushing them back inside. When all the intestines were back inside, we tightened the leg rope so it partially rolled the horse on its back and exposed the opening better. There was only the owner and I alone so we had to improvise.

This time I went up into the hole with a half circle needle and a long piece of catgut and sutured the edge down to the leg side of the inguinal ring area. I didn't know whether the stitches would hold when the horse got up because there wasn't very much of anything to sew to, but I'd done the best I could and the horse was waking up. The owner and I held the horse's head down as long as we could to let him be as awake as possible when he made the effort to get up. I didn't want him to struggle and flounder around and fall several times getting up. Finally, I said, "You pull straight forward on his head and I'll get his tail and maybe we can keep him on his feet." He rolled up and got his front feet out, then slowly pushed himself up. We were able to balance him until he could stand on his own. I felt up the flank and everything seemed to hold.

I explained to the owner that there was a good chance he would die from peritonitis even though we'd given him some sulfa drug in the vein. He said, "Well, he's made it this far. I thought he was a goner for sure when those guts came out, but we got them back in and he's on his feet. Maybe he'll make it." I checked by telephone each day and about a week later I stopped by to see him. The horse was eating well and getting around as if nothing had happened. From that time on I chucked the pack idea and sutured them shut when I operated, but each case always caused me some worry.

On Trial

One experience after another gradually improved my image and proved my worth, but for many years, in fact, all down through the years, I've felt that I was on trial. It has seemed to me that veterinary medicine has been on trial, and I have felt a strong obligation to prove its worth and gain respect for the profession. It has been a constant incentive for me to do my best and help with all the animals in whatever way I could.

One episode with pigs points out how I was tested. It went like this.

"Basin Vet Clinic, can I help you?"

"Is this Dr. Dan?" The voice was a little familiar.

"Yes, this is me."

"Dr. Dan this is Clyde Killian, and I've had two pigs die—one yesterday and one this morning. Can you come out and see what's killing them?" That's who it was, Clyde Killian. We had gone to high school together and he had married Joy Frandsen. Joy and I had been classmates in elementary school as well. She had been a cute girl and sometimes my secret crushes had included her. I was so bashful then that I kept them plenty secret.

"I can come in an hour or so. Did you notice them sick or just find them dead?" Most pig diseases that caused death usually showed symptoms for a day or so before the pig would die.

"I never saw them sick at all, just found them dead in the pen." That told me something.

"They are penned up then, and you know what's been fed to them. Have you changed feed in any way?" Maybe he would give me a clue. Feed changes often cause trouble.

"Nope, they been fed just the same. We give them some fresh-cut alfalfa every day along with grain, free choice, and they have water in the pen. Somethin's killing 'em and I can't figure out what it is. Must be some new disease that's hit us." The weather wasn't hot enough for heat prostration. What else could cause sudden death? I wondered.

"You say they have water all the time?" I wanted to be sure.

"Yeah, they even have a mud hole to waller in and cool off."

"It sure doesn't ring any bells with me. I'll be out and open one up as soon as I can make it." There were times when I sure wished that laboratory help was closer. My mind began to churn out the possibilities. Dying suddenly like that sounded like poisoning, but they were locked up and there had been no changes in the feeding pattern. I wondered if birds or cats had tipped over some insecticide. I'll have to look close for something like that, I thought. I arrived and began looking and asking more questions. The insecticide was ruled out, but no more clues surfaced. I got my knife and started on the necropsy, trying with all my mental capacity to think and not miss something. The skin and muscles looked normal, and the blood was a normal color. I carefully opened the abdominal cavity and took time to observe everything there. The only thing that I thought seemed off a little was the stomach. The lining appeared a little inflamed, but I wasn't that sure. At least I didn't feel it was too significant, but it was the only thing I found. I opened the chest cavity and examined the lungs and the heart. Nothing I found was adding up to death causes. I was silently thinking as deep and far back into my classes as possible.

"Can you see anything yet, Dr. Dan?" I think it was Joy who asked the question.

"Boy, not very much, the stomach is a little reddened, and the history of sudden death without any other pathology says it's got to be some kind of poisoning, but I can't figure what it could be." I was stumped and I wanted to be honest with them. I was about to suggest that we try getting some samples to the state lab.

"Well, you're right. We figured out what it was before you got here, but we wanted to see if you could figure it out. We sprayed the apple trees and the hay we'd been cutting is under them. The spray must have settled on the hay." Clyde explained, a little mischievously I thought.

First Impression

It was probably my third spring of practice in the Basin when I received a call from Mr. Clinton Bowden from Altonah.

"Basin Vet Clinic, can I help you?"

"Is this the doc, the vetinary?" I knew he was somebody new.

"Sure is; Dr Dennis speaking."

"This is Clint Bowden. I've got a cow that hasn't cleaned. It's been two days and she's down and can't get up. Could you come up and take her afterbirth an' treat 'er?"

"I can come in about a half hour as soon as I get free. Where do you live?"

"Altonah, or I should say a half mile north and two miles west of Altonah. It's on the corner where the road turns."

"Okay, I'll be on my way as soon as I get free." I hung up the phone and about that time I realized that I hadn't asked him any particulars about the cow. I should have asked for more information. As I began to piece together what he had said I realized three things: she had just recently calved, she couldn't get up, and she had a retained placenta. I wondered if it were a milk cow with hypocalcemia, or a nerve paralysis, or maybe a broken leg. If it were the milk fever I knew I could be of significant help, but if it was the other two it was a matter of diagnosis, and then telling him what had to be done.

I found the place and pulled in past a milking barn and a corral full of Holstein cows. He was waiting for me at a gate. My hopes of it being a milk fever case brightened a little.

"She's up in the field in a stackyard where I pulled her calf. Follow me, I'll ride my horse." As he opened the gate and I pulled through, I thought he must be related to the Bowdens in Myton, because I thought I could see a resemblance. As he led me up the side of the field, I began to put some more pieces together, thinking that if he had to pull the calf she was probably paralyzed—an obturator paralysis where the nerve innervating the inner muscles of the leg gets pinched between the bones of the calf and the bones of the birth canal. There's no successful treatment for that condition.

We crossed a ditch and headed east across the top of the field while my mind still churned. I should have asked him more about the cow before I came, because it might have saved a trip, but he was concerned that she hadn't cleaned. I can remove the placenta. That seemed to be his greatest concern when he called. Then I'll just have to tell him what I have to tell him.

As we pulled up to the stackyard, I could see that it was a first-calf heifer and she was lying flat out. She didn't try to roll up on her brisket when we approached like one would that had any chance of recovery. The dead calf lay over by the fence and Clint told me he'd had a very difficult time getting it out. He said the head was big and he had to work for a long time to pull it through. I knelt down with my jug of water and began to wash her up and she didn't even move an ear. She was depressed and didn't have much will to live. I expected to find a lot of tears and trauma to the birth canal, but Clint had done a good job in that respect.

"Here, can you hold this handle?" I said as I held up the tail for him. Sometimes their reaction to being that close to the muck and smell could tell me a little about a person.

"I'll do anything you can get me to." He smiled. I began to explore deeper inside.

"Are you related to the Bowdens at Myton?" I asked trying to keep both of our minds off the ripe odor coming out of the heifer.

"Yeah, he's my brother. How'd you know him?"

"Well, I come from Myton originally; I grew up there," I said as I worked to peel loose the placenta from the cotyledons inside. I was in up to my shoulder and the odor that close to my face wasn't too pleasant, so I remained quiet for a time. The heifer seemed extra warm like she had a fever, but the placenta soon came free and I pulled it out. Clint heaved a big sigh of relief like we were home free. When I went back in to feel if I had removed everything, I found a big tear into the abdominal cavity. With that and the fever a serious flag flew up in my head. I knew he probably caused the tear by pulling on the legs without the head being started, but I didn't think it would make him feel very good to tell him. It was a very difficult moment for me. I knew in one way I should explain how it happened and that a C-Section could have saved both the cow and the calf, but that would be censoring him when he had done his best. I knew the ranchers and farmers were self sufficient and did what had to be done because of necessity. There had never been any help except a neighbor and they just did the best they could and lived with the results, yet I needed to teach them somehow. It was a difficult decision, but I resolved to not embarrass him right then. I took my jug of water and walked back over to my truck and pulled off my coveralls and began to wash my arm and shirt sleeves where the juices had run up my arm and down inside my shirt.

As I was washing up, I noticed a lane at the end of the field which led down to the main road. I thought it would be closer than going all the way back around the way we had come. When I had everything back in my truck and looked around like I was ready to go, Clint pulled out his check book.

"What do I owe you?" My mind quickly tallied up the mileage and what I'd done, then I couldn't help but factor in that the heifer was going to die. I ended up with about what my expenses would be and told him. He wrote out the check, tore it off and handed it to me.

"Now, what do you want me to do for her?" He said pointing to the still prostrate heifer. All at once I realized that I had to bite the bullet and level with him.

"I don't think you'll have to do anything. I think she'll be dead in the morning." I said rather point blank thinking he would ask me why and I could explain. He just clammed up and didn't say a word. After a time I thought I ought to break the silence.

"Does that lane over there go down to the main road?" I asked pointing over toward a gate. "It would be closer than going back around wouldn't it?" He hesitated for a moment.

"Yeah, it's shorter," was all he said. I climbed into my truck and pulled over to the gate. It was dark enough so I turned on my lights, got out and started to undo the wire on the gate. Clint rode over on his horse. "Maybe you better not go that way, you might get stuck." I didn't want to get stuck especially in the dark so I did up the wire and we went back the way we had come. When we got back to the corrals he stepped down and opened the gate. I waved a greeting as I went through the gate and headed for home.

It was probably six months later when Clint stopped in at the clinic to ask about shots for his dog. He said I was right, the heifer was dead the next morning. I never did tell him about the tear in the uterus, but over the years he called me many times and we learned together when I could help and when I couldn't. He became one of the finest friends I ever had. Besides having a dairy herd and beef cattle, he and his wife had a soft spot in their hearts for small companion dogs. Down through the years they had several dogs and I was called upon to treat, immunize, and operate on them.

Toward the end of his life Clint stopped at the clinic to ask some questions about his dog. He usually called ahead to see that I was going to be there, but this day he just stopped in. I wasn't around and he began to reminisce with Dr. Whiting about our relationship over the years. It was then that we learned the rest of the story about my first call up to his place. He told Dr. Whiting all about what had we done and then about my comment that I thought the heifer would be dead in the morning. He said, "When he made that statement, it really hit me crosswise, and I thought, you young smart aleck, what kind of an answer is that? I was still smarting under the collar when he asked me about going down that lane. I knew it was so boggy down that lane that you could swamp a saddle blanket, but I thought it would serve that smart aleck right to get stuck, so I told him it went down to the main road. By the time I got on my horse, I decided I'd better not let him go that way. It just wouldn't be right, and I'd have to get my tractor and pull him out anyway. I rode over and told him he better not go that way."

5
REGULATORY WORK

Federal Regulatory Start Up

I hadn't been in Roosevelt very long when the federal veterinarians, Dr. Jones and Dr. Winward, dropped in to see me. I was acquainted with them from family contacts I had made while at Utah State and also when they had come to our farm testing for tuberculosis in earlier years. They had been working in this part of the state on the TB and Bangs control program for many years. In order to maintain a modified accredited TB status, a percentage of the cattle needed to be tested. They were doing that testing and also maintaining several purebred beef cattle herds on a brucellosis-free status so their cattle could be shipped to other states. They were very glad that I was here, and happier still that I might be able to take over some of the work. In fact, they were anxious to give it all to me.

It was at a point in time when there was a lot of pressure to try to complete the eradication of brucellosis. This would entail testing every herd in the area. We practicing veterinarians did the work on a per head fee basis, but the scheduled amounts were very low. From one to ten head had a trip cost figured into it, but from there on you couldn't much more than make your expenses. If there were any other vet work to do, you couldn't afford to miss it for the Bangs and TB work. I did what I could for the first year or so and did quite a lot of testing at the livestock auction. I had to run an agglutination plate test while the owners waited before I could release the cattle back to the farm. From the work I was doing, I could see that there was no way I could hope to do all the work in the two counties even if they paid more. In addition, calfhood vaccination of the heifer calves was getting into full swing. They kept encouraging me and twisting my arm, and when I went to the annual State Veterinary Association meeting in Salt Lake, the Federal Supervisor, Dr. Melvin, asked me to meet with him in his office. When the afternoon break occurred, I went to his office and he informed me that the Wasatch Front counties were moving along fine with the eradication program, but the outlying areas were lagging behind and he sure wanted me to get Duchesne, Uintah, and maybe Dagget Counties done. It was easy to understand why the outlying areas were lagging because the cattle were scattered far and wide, and there were few veterinarians.

I had engaged a couple of men to go along and help me on a few occasions, since much of the testing and about all of the vaccinating of calves was done on the end of a long rope. I had toyed with the idea of teaching them to do the bleeding and vaccinating and getting permission to have them help me. I knew that federal accreditation was a requirement to do the work, but I just didn't have the time to do it all alone. I was trying to build up my reputation and client confidence and I wanted to be available when they needed me, not out to Tim-buctu bleeding cows. I thought it was time to lay my cards on the table, so I proposed the idea of hiring the men, training them, then supervising them. Dr. Melvin about fell off his chair as he exclaimed, “Oh, no. We couldn’t do that, it’s strictly against regulation, and you better not do it, or you’ll loose your accreditation.”

“Well, I realize it’s against regulation, but it’s the only way I can see to get the job done. The fellows I have in mind are good, honest, responsible men, and I know I can train them to pull a tube of blood and put in an ear tag.”

“I can see your point, but I just wouldn’t dare do it.”

“Then you better scare up some more help because I know I can’t hope to get it done by myself. I think it will be impossible, but I’ll continue to do all I can.”

“We sure do appreciate what you’re doing. I was just hoping you could do it. All my men are tied up in other areas. Keep up the good work.”

“I’ll try.” The next day at the end of the meetings he asked me to stop in at the office again. His office was close by so I hurried over right away in order to get on my way home as soon as possible.

“You know, I’ve been thinking this thing over since you talked to me yesterday. It’s strictly against regulation, but I’ll tell you what I’ll do. If all the work comes in over your signature, as far as I’m concerned you’ve done it. Now don’t breath a word to a soul, and I’ll just turn my head. You realize if we get caught, I’ll have to deny any knowledge of it whatsoever. You’ll be on your own and probably in a lot of trouble.”

“Well, I’ll have to do all the scheduling and the paper work anyway, and I’ll be supervising it on a daily basis. I think as far as the owners are concerned they’ll think I’m the one doing it, just using some help. I’ll have to check out and brand the reactors and write any permits so I think I’ll have a good handle on it.”

“All right. Good luck, but don’t you tell a damn soul I knew anything about it, okay?”

"Okay, mum's the word, thanks. We'll do a good job."

The work started forward and I burned a lot of gasoline and midnight oil to keep up with it. When they brought in the blood and vaccination records I'd have to fill out the official forms and get the blood ready to mail. There were some late hours and much time consumed keeping vaccine ordered, herds lined up and suspects and reactors checked out. It went on for the next ten years or more before we got the job done.

Right in the middle of the process some tuberculosis showed up in some dairy cows from this area and we had to test about fifteen to twenty thousand head of cattle to re-establish our modified accredited status again. Since this was tuberculosis testing, only veterinarians could do it and I didn't get a lot of help from the Feds. I think my main problem was reluctant owners who'd just done the Bangs test. They didn't want to handle their cattle again so soon.

It was a long and tedious process to get all the herds in the state tested, and it required some changes in the laws and regulations to bring some owners into compliance. It was about midway, or a little farther down the road when Hugh Colton, serving as president of the State Cattleman's Association, asked me if I would accept an appointment as chairman of the Animal Health Committee. Having a small herd of cattle myself, I had become very active in the Duchesne Cattleman's Association, serving as a director and then as president where I had worked with Hugh. It was a very critical time and I was influential in obtaining the support and commitment of the Cattleman's Organization and bringing the Utah Veterinary Medical Association and the regulatory agencies together to get the cooperation necessary to finish the job. I felt that I had contributed significant leadership and wisdom that would help the livestock industry and the whole state of Utah beyond my lifetime. It was an especially rewarding experience.

Testing The Wild Ones

The testing of the smaller herds that were pasture raised

and those that went to the mountain and returned to the ranches for wintering were not too hard to schedule and get done, except everyone wanted to do them at the same time, when the calves were sold in the fall. It was the bigger herds that remained scattered on the range all year that posed the greatest problem. A few examples may show the challenge.

I had talked to Willis Stevens about doing his herd out in the Willow Creek area. He told me they never corraled all of their cows at once and about the only reason they ever got them in was to pull the calves off. This was done in several range corrals without chutes and facilities. The one good thing about the regulations was that only twenty percent of any one herd had to be tested and if no infection showed in that many, the rest were considered clean. I explained this to him and he said he'd work out something and get back in touch.

Like most owners, if he were going to test, he wanted to test as many as possible and not just the minimum. He had a lot of very wild cows, but he came up with an ingenious way in which he thought we could capture most of them. It was in February when he called me up to set a date. I got the tubes, tags, needles, clipboard, papers, paint and brands all ready the night before. It took a long time to number ten boxes of tubes and get everything ready for a 5:30 A.M. takeoff, then hit the hay at about 11:30 P.M. We used paint brand numbers on each cow in sequence with the last numbers of the ear tag. This way a man on a horse or in a truck could pick out any that were reactors or suspects without having to round up the herd and put them through a chute to read ear tag numbers in case any tested positive. He told me to be there between 8:30 and 9:00, in the morning, so I figured I'd better be on the road by 6:00 in order to make it and give me time for a flat tire, which wasn't uncommon on those roads.

I left home a few minutes after six and it was still dark, but morning was sure to come. There weren't many people out at that hour during the cold winter months, but by the time I got to LaRue Pickup's place below Randlett he was just leaving the house to go milk. I honked as I went by and he waved. By the time I got to Ouray, it was starting to get light a little and smoke was coming from a home or two. The dirt road from there on was in good shape with most of the snow gone and the mud frozen solid. I watched the sun come up several times as I went up and down the ridges between the Turkey Trail Dugway and the head of Buck Canyon. It was late winter and there were deer all along the route through the cedars. When I started down Buck Canyon, there was a herd of them and I about missed the road in a place or two trying to count and estimate their number.

When I arrived at Willis's house, his cowboys were saddling up some horses, and he was putting some sacks of protein pellets in the cab of his pickup. He was opening the sacks and arranging them so he could reach into each one. The three cowboys loaded the horses in the big truck and hauled them down the canyon where they unloaded and each rode up a separate draw out of sight of the cattle. Willis told me to stay back until the cowboys were in behind the cows then follow along. You see, he had been feeding those protein pellets to his cows down there in the bottom of the canyon all winter in preparation for this capture. The cows eat those pellets like kids do candy and literally fight for them. He had a big corral down the canyon with some fenced wings that came out from the gate, and that was his destination.

He started down the road pounding on the side of his truck and calling to the cows while he threw out a few pellets. Those cows came down from the ridges, out of the draws, across the creek and came running to that old pickup truck from everywhere. I hadn't been able to see very many, but there were about 350 or more crowding around and following that truck. Willis kept tossing pellets out as he drove and the cows cleaned them up as clean as a pin. He drove down the canyon and right up the slope into that big round corral with the cowboys spread out behind him. As he entered the gate he pulled a sack over to the window and poured out the goodies in a circle as he drove around the perimeter and back to the gate. The cowboys and I closed in on the herd as it neared the corral and began to push the tailenders as fast as we could. When some of those old wild sisters realized what was going on, they wheeled around and made for that gate nearly knocking the horses down on their way out. There were only a half dozen or so that escaped before Willis got the gate closed, but there was no way we were going to stop them. I realized what he had meant when he said the timing would be right about 9:00 A.M. That was the usual time he fed pellets and the best time to try the capture plan. That half dozen that ran over us went across that canyon and were still running when they disappeared over the far ridge. This was the pattern that a lot of them followed as we turned them loose from the chute after pulling their blood samples. It was a long day putting nearly 400 head through that chute to get a blood sample, then tagging and branding them, but it was still daylight when I started home.

I went down the canyon past some of the other ranches and the old school house. My uncle Efe Birchell's old house was crumbling and his corrals had mostly fallen down, but I recalled how he cherished the dances they held in the old school house with a tapping boot and a fiddle. As I climbed up the hill, I tried to recall the stories my mother had told me about how the dugway got its name of the "Turkey Trail," and how she had ridden with Uncle Hank Stewart to deliver the mail. Someone had taken his team and hand plow and wound around the little draws and ridges to make a wagon road down into the canyon. The neighbors accused him of turning a turkey loose and following it as it went down to get a drink. The mail delivery went up Hill Creek, across the top and down Willow Creek taking about five days to make the trip with a pack string. I had a lot of memories brought to mind of stories passed down to me from an earlier generation. The time passed quickly for me as I traveled along. When I passed LaRue Pickup's place, the lights were on again in the barn and it was way past dark when I got home, but it had been a good day—one that I enjoyed.

When I picked up the mail with the test results, I was

nearly holding my breath. I'd thought many times in the three intervening days what we would do if he had some reactors. How would we ever get close enough again to read the paint brand numbers, and if we could find them, how would we get them corralled again? I thought Willis might just shoot them with his rifle if he could tell for sure which ones they were and butcher them out like a deer. Fortunately, every one of his cows were negative and no reactors were ever found out that way. I think the dryness of the area combined with sub-zero winters and the wide scattering of the cattle prevented the spread of all infectious diseases in this and many other areas of the Basin.

Another memorable trip was going to Nine Mile Canyon to do the Preston Nutter herd. This had to be arranged in late spring when the cattle were moving from the winter to the summer range. They were gathered to the home ranch before being pushed out to the south summer range. The herd was divided and they didn't want to mix the two groups so we had to handle them on two different days. This necessitated my staying overnight, resulting in one of the most interesting evenings I have ever had. They had me sleep in a guest room in the owner's home instead of the bunk house where I expected to sleep. Howard and Virginia Price were running the ranch at this particular time and we spent a long enjoyable evening talking about Virginia's father, Mr. Preston Nutter, one of the most famous ranchers of the old west. His old trunk was there in my room and they let me browse through seventy-five years of papers and treasures that he had accumulated. His old pistol with the notches on the handle was there and many other mementos, along with a lot of old papers. I picked up one three cornered piece of brown paper that appeared to have been torn from a paper sack. On it was written a letter to one of his foremen in southern Utah. The message, as I remember it, was to round-up and tally brand 5,000 steers and deliver them to a certain railhead on the last of September. At the bottom was his signature. I got to see and talk about a lot of his life as I spent that evening with Howard and Virginia. Howard sat with a cold pack on his head from a bump he received during the first day of testing.

Howard was a retired army colonel who had never been around cattle in his life, and this was early in his tenure as manager of the Nutter spread. He brought many of his military ways with him to the ranch. Things had to be done just like he wanted it, and many times this wasn't the most practical way when handling wild range cattle. There was a large pen full of cattle behind a small catch pen and a long chute where we were bleeding, tagging, and paint branding. As we finished with each cow she was released into a long pen up against the high rock wall of the canyon. This pen connected through a gate to the big pen full of cattle to be worked on. When you handled some of those cows against their will they weren't very happy about it and some would turn to fight before retreating to the far end when they were released. We had put about a dozen through the chute and they had retreated to the far end after blowing a little snot at us to let us know they were mad. We watched attentively as each went out because these cattle were not dehorned and carried some formidable, long, sharp horns and we didn't want to tangle with any of them.

At any rate, Howard noticed something about the way the cowboys were pushing cows into the catch pen that wasn't just like he thought it ought to be done. In his typical military manner of violent hand signals and commanding voice, he marched down through the long pen hollering and waving his arm to get the attention of the men. He wasn't paying any attention to the cows, but you can well imagine the attention he was generating in some of those upset ones we had released. Several of them began to paw the ground and came right out to meet him so we began to scream out a warning. By the time he noticed the mad cow coming at him it was too late to get to the fence so his only option was to dive under an overhanging ledge. There was just room enough for him to roll up tight in the low space under the rock where the old gal's horns couldn't get at him. That snotty old cow was hooking and raking that old cliff, making smoke fly off the tips of her horns as she tried to get to him. The whole operation stopped as we teased and coaxed the cows away so he could be rescued.

When he made that first dive he had hit his head on the rock hard enough to daze him, but luckily not enough to knock him out. After he went up to the house to recover, I heard the comment, "Howard just don't realize that some of these old cows outrank him."

One more testing experience with the wilder range cattle has left a lasting memory. There was a herd of about two hundred up in the Strawberry River area. By the time I arrived all was ready to go. The cows had been corralled the night before. There was a large rectangular pen on the east side connected by a gate to a smaller square pen on the west. From this pen the cattle were pushed into a small catch pen and down the chute. The gate connecting the two larger pens was against the north fence and had been left open during the night allowing a goodly number to drift over into it. When the owner and helper arrived they had cut off about fifteen more and pushed them into the square pen giving us a good bunch to start on. Things were moving along very well although the cows did a lot of bellering as I poked needles in their necks. This stirred up and made the remaining herd very nervous since they were just across the fence from me.

When we finished that first bunch, we all climbed over into the big pen to push some more into the square pen so we could work them. One man opened the gate and we began to move the herd up toward it, but the cows were stirred up and suspicious. We couldn't get them to move and especially not through that gate. They all put their heads toward the middle of the herd and just pushed. We had some long willows and prods and we kept pushing them up toward the north fence. When we had them crowded up against the north fence, the hired man started to cut off a package to push through the gate and they began to shift their pushing toward the fence. The old posts gave way and that whole north fence fell flat. Those cows went up over the hills and ledges in every direction like a scared herd of deer. Within minutes you couldn't see a cow.

I retreated to my truck and counted up the blood samples—we had forty two. That was two more than the minimum twenty percent needed. I explained to the owner that if they were all negative we wouldn't have to test any more. He said it didn't matter much because if they didn't pass he couldn't hope to get them in again until next year at about the same time. Once again it was a brucellosis free herd, and once again I witnessed how much pressure a herd of cows can exert on a fence.

6
INVESTMENT IN THE FUTURE

The Roots Begin To Grow

The old home on Main Street left a lot to be desired and, once it looked like we might get enough business to stay alive, we began to think of a home of our own. At about the same time, Max Owens decided to move to Salt Lake and he wanted to sell his home. We were favorably impressed, especially with how well built it was. He had constructed it and had really gone the extra mile to do a good job. It had a basement apartment where the kitchen and the living room could be rapidly converted to an office and small animal treatment and surgery room, leaving us with two nice bedrooms, a bath and a storage room. It seemed ideal so we applied for a G. I. Loan and purchased it on a twenty-year contract for $13,500. We were thrilled to have our own home even though we would be twenty years paying off the mortgage. It made a good working arrangement for our practice. I would do whatever surgery came along at night because I was out during the day or over helping on the farm so the big kitchen window became the TV screen of that day for all the neighbor kids and passers by. Sometimes they were three deep out there looking in on the operating table situated lengthwise right in front of the window. Many times the parents, coming to get the kids home, stopped and watched as well. I could have pulled the blind but they were so fascinated and caught up in it I didn't have the heart to do it. Out in the backyard they would watch me deliver calves in trucks, sew up vaginal prolapses, perform C-sections, and do rectals on both horses and cows. I think sometimes they got exposed to life's realities a little early and had many tales to tell their parents, but I didn't get many complaints.

Bear Amputation

One time I got a call from some fish and wildlife officers who had a black bear that had been in a trap and needed a front paw amputated. I told them to bring it in and of course my family was all excited. No one was more excited than myself because I'd never worked on a bear, and I didn't know how I was going to handle the situation. They were about an hour and a half out when they called and my kids mentioned to some of the neighbors that a bear was coming in, so we had kids coming out of the woodwork by the time it arrived.

When it did arrived it was tied in a big heavy burlap sack like they put Gilsonite in. I felt around and poked him a little to see how wild he was going to be and he growled, but didn't fight much. He only weighed about a hundred and fifty to a hundred seventy pounds being just a young one, so I calculated about how much anesthetic he would need and put it in a big syringe. Then I took a long needle and by maneuvering the sack I could get the needle up against his abdomen. I kinda jabbed him in the side with my finger until he quit moving then gently pushed the needle through the sack and into his abdominal cavity in front of the hip. I shot in the anesthetic and waited to see the reaction. He was soon relaxed and seemed to be out of it, so I had them carry him down the stairs into my surgery room. We took him out of the sack and put him on the table. I thought to myself "what if he suddenly comes around and you have a loose bear in your home?" so I watched his reactions very closely. By the time we took him downstairs, not only the neighborhood kids, but many of their parents were there to watch. I let a few come down and stand against the wall, but the rest had to watch through the window. The surgery went well with Joyce and the two officers helping, and by the time I cut off the bone and sutured up the end he was beginning to come out of the anesthetic. We carried him back up the stairs and out to a cage in their truck, and they took him to a zoo to recover. It was quite an experience, one which was talked about for a long time. I became a hero, and a brave one at that, in the eyes of the kids.

Entering Cow Business

A few years after we moved here my father and I felt the need to better utilize the farm we had by running some cattle instead of just selling the hay and grain. At about the same time I learned that Mr. Jim Lemon wanted to sell his permit on Anthro Mountain, which was out on our side of the Basin. He had a sixty cow permit and in order to sell it the forest service required the sale of cattle to go with the permit. We talked to him and felt his price was fair and he had good quality cows that were used to the range. It seemed like an answer to our prayers in one way and like a dream fulfilled in another. It gave our lives a new purpose and an opportunity to share it with each other and the whole family. What a privilege it was to see the new calves in the spring, to ride together on the mountain, to camp under the stars at night, and bring the herd home in the fall. It gave dad a new lease on life and although as the years progressed it demanded more of my time than I had to spend, it was very fulfilling. I learned that all the cows don't come home in the fall, that it's a cold and scarry ride to go to the mountain in the snow to get one. I found that we still have cattle rustlers who operate with trucks and trailers. I learned a lot about trying to drive one-eyed cows, the problems of cows not used to the range, and a hundred other things.

I even learned a couple of things pertaining to my profession. One was a cow that wouldn't eat much and began showing an abdominal distension. I thought that it was heart trouble, but her heart sounded normal. Hydrops-amnii came to mind, but it wasn't the right time for that and a rectal exam pretty well ruled that out. I punched at her sides, checked temperatures which ran about one degree high and finally determined that her first stomach, the rumen, was just filled with a huge amount of water. I put my big Kingman stomach tube down into her rumen and siphoned off what appeared to be a whole barrel of water. She was slab sided and gaunt when I got through but she could walk more normally and seemed relieved. I turned her out and returned the next day to find her filled up again. In the next two weeks I'd return to the ranch and empty her out when I could, but she only got weaker. She seemed dehydrated and acted like she was dying of thirst. She would go immediately to the water and drink when I emptied her stomach. It was a very puzzling case and one no one could have afforded to treat as much as I treated her but she was my own so I went on trying. Finally it became apparent that she was eventually going to die so I decided to do a necropsy to see if I could find the answer to the puzzle. When I opened her up the only thing abnormal was the pyloric end of the abomasum, the fourth stomach. It was inflamed and swollen to a thickness about six to eight times normal from about six inches inside the stomach to about the same distance down the duodenum, the gut. As I carefully sliced open this area there was one small strand of fine screen wire about one inch long that was wedged crossway so that the ends just penetrated the sides. I was amazed that it had remained there and irritated those linings enough to make them swell up that much. The result was that the pyloric reflex which pinches off and forces ingesta into the gut had been effectively negated to the point where the cow starved to death. It was a one-in-a-million occurrence. I have never seen or heard of it again.

The other lesson was an accidental learning experience which many subsequent cases have benefitted from. During the calving season dad or I checked the herd each afternoon to pick out any heifers that might calve, and take them to the corral where we could watch them. One day, like most times, we both checked the herd but we missed one that tried to calve during the night. The next morning I hurriedly pulled the dead calf, but the heifer was paralyzed and couldn't get up. She was up on a road embankment so we backed the truck up to it and slid her in, taking her up to the corral where she could be fed and watered without the other cows stealing her feed. We had a narrow loading chute coming off from the working area of the squeeze chute and there was a gate across at that point. We backed up to the unloading chute and pulled the heifer out of the truck and part way down the chute. At this point she began to struggle to get up and I helped her by lifting on the tail. Once on her feet, when she started slithering sideways, her hips hit the side of the chute and she could stay up. Having watched hundreds of them slither around and fail to stand out in the open, this was a revelation to me. We left her penned in this narrow alley with feed and water for a couple of days and she learned how to get up and down and use her legs. When we turned her out she got along just fine. This then became my recommendation to all owners when obturator paralysis from calving occurred. I told them to take them to a narrow alley or build one around them with panels, and I have found it has been a far superior treatment to anything else, and I learned it from one of my own cows.

Chuck's Three Legged Dog

I think it could be said that if Roosevelt, Utah had a

trademark it would be service stations. They sat in pairs or three's on almost every corner of Main Street. I used the one on the west side as the highway topped the hill coming into town from the west. Right next to it on the north was a brick home where Charles Edwards lived for a number of years. He had a fair-sized shepherd mix dog that loved to fight. With many farm vehicles carrying dogs in the back stopping at the station, and the dogs on the south end of town coming by, he had ample opportunity. If he could arrange an encounter he always did and seldom, if ever, came away the loser. It caused Chuck and his wife a lot of embarrassment and quite a number of complaints. Keeping him tied up was the only control and that didn't always work, because some dogs would wander in to challenge him. The chain on his leash was then an unfair handicap, but it didn't deter him from his fun.

One day Chuck showed up at my place with the dog. One front paw was very swollen and enlarged and he was dragging it. The skin on the front was all worn off and he had a big, open sore.

"Doc, we've been thinking maybe it's time to put our dog to sleep." Chuck and his wife were high school classmates. In fact I'd gone all the way from first grade on with him. I could tell it was a very painful decision and not altogether final.

"Looks like he's got a radial nerve paralysis and drags his leg. He's worn all the hide off." I could see the exposed parts of his foot.

"I guess it is; he can't pick it up. He got hit by a car. Didn't break anything that I could tell. I saw it happen and it didn't run over him, just knocked him for a roll, but ever since he's drug that leg." He reached down and patted the old dog who wagged his tail. I knew what the trouble was, I'd seen a few.

"That's the way it happens. You see the radial nerve comes out of the body and passes around the point of the shoulder and down to enervate the muscles that lift and straighten out the leg. If they get a hard enough blow on that nerve to damage it, then they have to drag the leg."

"Can you fix 'em?" There was hope in his voice. I think he

thought that if I knew all about it then I might be able to fix it.

"I can't fix the nerve, but I can amputate that leg and he'd be as good as new in a couple of weeks. You'd be surprised how good they get along." I didn't necessarily want to change his mind, only make him aware of what could be done.

"You mean you cut dogs' legs off?" The idea was a bit foreign to his thinking and background.

"I sure do, and cats and deer fawns, and even a bear once. Mostly I fix up a bad end left by a trap or a mowing machine, and sometimes this kind, when it can help them to get along better. I even take a toe off of a cow now and then when they have a really bad foot rot infection on one side. It helps them to get along better, like it would him." He was thinking about what I'd said.

"Gee, a three-legged dog. I don't know whether Murl could handle that." He paused for a moment, thinking. "He's always been such a good ol' lovable pet, but I've had a lot of complaints about him fighting. I thought maybe it would be just as well to solve both problems," Chuck said, a little dejected. I didn't say any more, I could see he was thinking. A half smile stole across his face. "You know, since he hurt that leg I haven't seen him in a fight. Maybe taking that leg off would stop him from fighting." I could see his mind was changing.

"I don't know about the fighting. I do know that he would get along a lot better." I didn't have any idea how it might affect his fighting. I thought it should make some difference.

"I think I'll do it, Doc. Maybe with only three legs he might get licked a time or two and decide to mind his own business. Can you do it if I leave him?"

"I won't be able to operate on him until in the morning, but we can put him in a kennel in the garage. He needs to be off feed overnight anyway." We put the old guy in the kennel and the next morning I cut off his leg above the elbow and sewed it all up. He healed fine and in about two weeks I took out the stitches. After the hair grew all back I'd see him as I went by or stopped for gas and you could hardly tell he was missing a leg. It didn't seem to slow him down a bit, and as for his fighting—Chuck said in six weeks he was back at it and whipping everything that came along. It was one less leg the other dogs could bite!

The Turtle Episode

"Basin Veterinary Clinic, can I help you?"

"Is this Dr. Dennis?"

"Yes, this is Dr. Dan, what can I do for you?"

"Oh, Dr. Dennis, our turtle is hurt. He got run over, and it broke his shell, and he's bleeding. Could you look at him?"

"Sure, I'll be right here at my office. Do you know where I live?"

"Yeah, we know. I hope you can fix him."

My mind quickly raced back to when I was base veterinarian at Ethan Allen Air Force Base and my first experience with turtles. I received an anguished call from a young mother who explained that her young son had just dropped his turtle into some boiling water and she wanted to know what she could do. My immediate response—knowing that the turtle was already dead—was to facetiously suggest that she could go ahead and make turtle soup, but my better judgement took over, thank heaven, and I fought back the initial temptation and kindly explained that I thought the turtle must have died instantly and there was nothing anyone could do.

While the latest patient was coming, I got out some bone plates and screws thinking that might work on the shell. I wondered how you could determine if bones were broken or how badly the turtle might be injured. It wasn't my first turtle patient, but what did I know about turtles? We certainly hadn't studied anything about them in school. It was another one of the challenges that came along and I knew I'd have to rely on my own knowledge and ability. There wasn't anyone I could turn to for help.

When they came in with the turtle wrapped in a bloody towel I said, "put it there on the table, and let me have a look." I carefully took the towel away to discover a big crack in the upper part of the shell that went more than halfway around. The bottom of the shell was broken also. The body inside didn't seem damaged. There was a little blood seeping through the fractured areas but it was minimal and the wounds were not filled with dirt as one might expect. I carefully pressed the fractured shell edges back into place and wondered how much pain I was causing as the edges scraped together. It made an involuntary shudder sweep over my shoulders and back as I felt the crepitation between my hands, but the turtle didn't move or struggle. I just had a feeling that the old fellow was going to be all right. I looked around and all the kids were holding their breath.

"Are you sure you don't want to make a big pot of turtle soup?"

"NO!" rang out a chorus of voices and my smile told them I was just teasing.

"Well, if you're sure, we just might be able to fix him."

I had been weighing things in my mind and not knowing how screws through that shell might react, I gave that idea up for a new thought that came along. I thought if I could get that shell all back together it could be held that way and protected if I just put a plaster of paris cast all the way around him.

"We want him fixed, we love him."

"Okay, we'll clean him up and wash some of the blood off and disinfect him, then I'm going to put a cast on him just like when you get a broken leg or arm."

With the help of a couple of pairs of hands to hold things in place and a few rolls of plaster of paris we created a new white shell all over the old one. I worried about infection and swelling but the next day they said he was walking around. He apparently recovered because I never heard from them again. I've often wondered how long he wore that cast and how, if ever, they took it off of him.

Hardman Cow

"Dr. Dennis, how much would it cost for you to come over

to Myton to see a sick cow?" I knew some people in Myton but the voice didn't ring any bells.

"I charge about ten dollars plus whatever medicine we'd have to give her."

"Does it have to be paid right when you come, or could it be charged?" He sounded more concerned about the money than he was about the cow.

"Well I, ah, I'd like to be paid, but I guess it could be charged if necessary. Who is this?" I wondered about this kind of client.

"This is Hardman, Hi Hardman. I live down by Crapo's and our family cow is sick." I thought I knew the house where he lived.

"What is she acting like? Is she off her feed?" I asked, wondering if maybe I could get an idea of what might be wrong.

"Well, the first thing I noticed was that she dropped off on her milk to practically nothing, and she won't eat her grain. She noses the hay around and eats a bite or two—mostly the leaves. Whatever it is, it sure hit her all of a sudden. Yesterday morning she gave her usual twelve quarts, then last night it was only about half a bucket, and this morning she didn't give more 'n a cup or two. She's real sick and my kids need the milk, besides I sell some to the neighbors." The loss of milk and appetite sounded a lot like the hardware cows I'd seen in Idaho.

"Does she grunt, or walk a little stiff and humped up?"

"Yes, she does, she don't move around much." I was pretty sure what the old girl had. She'd picked up a nail or a piece of wire and it was poking through her stomach some place.

"Okay, I'll be over in about forty-five minutes. You say you live down by Crapo's, on the east end of town?" I wanted to be sure where it was just in case Crapo's had moved. It saved a little time and some embarrassment of having to ask where someone lived.

"We live in the house north of his."

"All right I'll be over as soon as I finish up some things

here." I had a good feel for what was wrong with the cow. It was a time when the pickup baler was used a lot and in the tying process, short pieces of the baling wire were cut off which found their way into the hay. Besides that, scattered wire and other debris were raked into the windrows and baled up in the hay. When the cow first eats her food, she merely wads it up and chews it enough to form a bolus she can swallow, and the wire and nails go down with the bolus. Once inside, the metal filters down through the feed to the bottom of the stomach where it tangles into the lining. Then, as it is sharpened by oxidation or rusting, it penetrates the stomach wall and causes a localized peritonitis accompanied by pain and reduced activity of the stomach. Back in Idaho, I had encountered it many times and had operated on about twenty cows or more. I put together my surgery kit with the procaine, antiseptics, and medicine I needed and headed for Myton.

As I drove along I thought, "Here goes another free ride. Why can't people see that I need money to feed my family and pay the bills?" I thought about the people who'd brought their dogs in and I'd checked them over and told them what was wrong and when I'd get through they'd say, "Thanks, Doc, does that cost anything?" They didn't seem to expect to pay anything unless I gave them some medicine. Even some that I treated acted like it ought to be free. I was frustrated and having a struggle to make ends meet, but I wanted to be friends and help people too. As I thought about it I decided that this time I'd check the cow and tell him what was wrong, but before I did the operation I was going to insist on being paid.

When I drove into the yard about eight kids came out of the bushes and the woodwork. Their clothes were threadbare and worn and they looked a little on the hungry side. They were bashful and didn't make much noise but their eyes were warm and intensely curious.

"Now you kids all get back into the house with the others," the father said as he shoved them like a flock of chickens. I made a mental note of his statement, "with the others," and wondered how many kids he had. I approached the old, gentle cow and quickly checked her over. She was indeed a very typical case of what we called "hardware disease."

"She's got hardware; that means she's eaten some wire or a nail and it's penetrating her stomach. We need to open her up and take it out."

"How much does that cost?" Here was his concern again about the money.

"Thirty-five dollars."

"What happens if she don't get operated on. Will she die?" I began to realize how much that cow meant to him.

"No, she won't die, but she won't get much better very fast. They usually improve a little in a few days, but she won't come back up on her milk much. You see, that wire, or whatever it is, pokes through the stomach and causes an infection and the stomach adheres to the body and doesn't function normally, so they don't eat and when they don't eat they don't give milk. Most of the wires settle and penetrate through the bottom of the second stomach. That second stomach is right against the diaphragm and the heart is right on the other side. If the penetration is toward the front, the constant beating of the heart turns the wire which works into the heart sack. Once the sack becomes infected, the cow will die, but it takes a few months for it to happen."

"How long does it take her to get over the operation if you do it?"

"Most of them are back on feed in about three days and the milk production goes right along with how well they're eating, but when you're dealing with living things nothing is guaranteed, you know." I didn't want to promise him a sure thing but I knew from experience that most cows came right back.

"Yeah, I understand. I don't know what I'd do without that cow. She about feeds my kids with the cream, butter and cottage cheese along with the milk they drink, and we sell milk to the neighbors too. I just don't know what I'm going to do." I could see that the poor old fellow was really having a hard time. I had already concluded that he didn't have any money, and he was having a hard time bringing himself to the point of asking me if I would operate and charge it. I thought if anybody ever really needed help, these good people did. My hard-nosed resolve on the way over melted away and I knew I couldn't leave without helping that old cow.

"Look, Mr. Hardman, why don't we go ahead and operate on the old gal, and we'll worry about the money later. I think she'll be right back on her milk in three or four days and you folks sure need it." The world seemed to slip off his shoulders and a light came on in his countenance. He swallowed a couple of times to push the lump out of his throat and blinked back some tears.

"Could you do that? I don't have any money right now and I'm not working. I don't know when I can pay you." I knew he was an honest man and I liked that.

"Well, let's not worry about that now. Where's the nearest power?"

"Right there in that shed." He said pointing over past the cow and moving with a lighter step.

"Okay, we'll tie her up there to that post and push her up against the wall. Do you have a pole we could wire up along the side of her?" She was a gentle old cow and didn't mind the pole against her. I got my kit and the clippers from the truck.

"Here, go in there and plug these in," I said to one of the boys that had silently approached the scene. I clipped the hair off the left side and began to wash the area with antiseptic soap. When I sat my jug down, I could see little faces peeking through almost every vantage point. When I started sticking the needle through the skin to deaden the area, the cow didn't appreciate that and she kicked at me with her back foot several times. Each time I could hear little snickers here and there. I was glad I had that pole to ward off that hind foot. When I had the side all deadened and painted with iodine, I took my scalpel and made a bold six-inch incision through the skin. I heard some small gasps and when I looked there were some faces covered with hands. As the blood dripped and I cut on down through the muscle into the abdominal cavity, it was totally silent. I caught hold of the stomach wall with some towel forceps and made the incision in it. When I had everything in place I said, "Here, Mr. Hardman, you hold these two, just remember we don't want any of the stomach juices to get down inside of her. I'll hold this bottom one with my left hand while I reach down inside and you pull those other two out and down over the outside."

When I reached down into the stomach, I had to force my hand down through a mass of hard contents to get to the bottom. There was liquid on the bottom and I soon found something hard poking through the stomach wall. I felt around the bottom of the stomach for any other foreign bodies, then pulled the long piece of metal out of the stomach wall and brought it out. It was a long finishing nail and when I held it up, little hands began to clap. I got my needle holders and a big curved needle with a long piece of catgut threaded through it. Once again there was silence as I closed the stomach and sewed up the hole in the old cow's side. When I clipped off the wire suture in the skin with my scissors and stepped back, she looked as good as new. There were whispers and bubbling voices all over the place, but none were as pleased as Mr. Hardman. When we turned the old cow loose he put his arm around her neck and gave her a squeeze, then patted her shoulder as she walked away. I thought to myself, "If I never get a penny for doing this it's been worth it anyway."

Two days later when I stopped by to check on the cow and give her another shot, I reflected on the feelings I had had when I operated and it still felt good to be able to help that family. When I got all through Mr. Hardman reached into his pocket and pulled out three tens and a five dollar bill. It was so unexpected and with my feelings I was hesitant to take it.

"I borrowed it from my neighbor," he said as he pressed it into my hand.

Sarcoid Surprise

"Dan, this is Parley Lambert. Could you come up to my ranch at Boneta and look at a horse?" Parley had been involved in various businesses most of his life, the most recent being an ice cream shop in Roosevelt. Most everyone knew Parley, and now he had acquired this ranch up in Boneta and in a sense retired.

"What's the matter with the horse?" I knew Parley didn't have a lot of ranching experience and I wondered why he wanted me to come clear up to Boneta to see a horse.

"Oh, he's got a sore on his leg." And that was all he told me. When I arrived he took me out to the corral and showed me this horse. It had a huge sarcoid, a cancer-like growth, all across the back of its foot. It was bigger than a softball and smelled terrible. It was about the first one I'd ever seen and I couldn't remember ever seeing one treated when I was in school.

"It's been on there a long time and it's growing, and when it's bumped, it bleeds like crazy. I guess that's why they call 'em blood warts. At least that's what my neighbor called it. What are we going to do about it, Dan?" He always called me Dan, we were sort of family. My brother had married his wife's younger sister. I was weighing things up in my mind. I couldn't remember any recommended treatment from school, but I figured it had to be cut off, and if it bled freely how was I going to stop the hemorrhage? I didn't have very much bandage material with me, and no chloral hydrate to put the horse to sleep with, and there I was thirty miles from the office.

"Maybe we could burn it off, Parley, and sear some of the bleeders as we did it." I really felt unsure being no more prepared than I was.

"Sounds good to me. What do you want me to do?" he said expectantly. "I've got a couple of good men lined up to help if we need them. They're just up the road, I can call them." Realizing again that I didn't have any anesthetic we'd sure need them to help throw the horse.

"You better give them a call. We'll have to throw him with ropes. Do you have any old car springs laying around?" I had a flash of how they were tapered and flattened on the end. I had recently finished putting an extra leaf in the back springs of my truck. They should work like a knife if they were hot. “We’ll need some wood to build a fire.” He called to Hazel to call the neighbors and then turned back to me.

“If were going to build a fire, we’ll need to get away from these corrals and hay stacks. How about out in that little pasture?” He pointed over south of the house.

“That looks good to me, what about the springs?” I was still visualizing that stinking mess on that horse’s foot and how I was going to get off. He soon returned with four good sized leaves from an old pickup truck. We put the wood and the irons in the back of my truck and I pulled around through the gate into the pasture. Parley came with a little can of diesel and we soon had a good fire going. When it burned down to some coals and the springs looked hot I told them to bring the horse. I put the throwing rope around his neck and we fished the two ropes between his hind legs and back up through the neck part, keeping the ropes above his hocks so he wouldn’t step out of them.

“You hold his head and I’ll take the rope on this side, you and Parley take the rope on the other side,” I instructed the two men and Parley. “Now I guess we’re ready. Work that rope down below his fetlock like this,” I said pushing slack back to let the rope slide down. “Now everybody pull,” I said as I heaved back on the rope I had. The man on the head couldn’t keep the horse from backing up. “Pull out sideways,” I shouted as we fought for awhile. Finally the horse laid over. I had the top rope, so I pulled the foot up to his body and figure eight-ed the rope around his hock and hoof, cinching it up so he couldn’t hurt himself. We then rolled him over and did the same for the other leg. I then grabbed my lariat from the truck and threw a figure eight knot into it and put it on the two front feet above the sarcoid. We worked the end under the horse and sucked his front legs up and tied them. We all heaved a sigh of relief. It hadn’t looked like we were going to get him down for awhile.

I wrapped an IV set around the leg to control hemorrhage, then I took a burlap sack and picked one of the springs out of the fire. It was good and hot and I sliced that bloody mess off the back of the foot. I got a fresh hot iron and cut down at the skin edge, dishing it out, to remove all the tissue that looked cancerous. The horse flinched a little once in awhile when I hit living tissue, but for the most part, he didn't seem to feel it much. I released the tourniquet and kept touching the bleeders with the hot iron until most of them had stopped. I powdered the whole area with sulfa drug and bandaged it as tight as I could with the small amount of bandage that I had. "Okay, let him up," I said as I undid the ropes and took them off. The horse rolled up and stood on his feet, and there was only a slow seeping of blood from the foot. It was a great relief and I felt good about the results.

"That was a good job, Dan, I think he'll be okay now, don't you?" Parley was as pleased as he could be.

"Leave that bandage on three days, Parley. Then cut it there at the knot and unwind it. Don't put anything on it, just let it heal by itself." I gathered up my stuff and left.

It was about six months later when I ran into Parley up town. "How did the old horse do?" I was almost afraid to ask him.

"Oh, he healed up fine. It took awhile, and he's got a scar, but otherwise he's as good as new."

That was one of the successful treatments of a sarcoid, but over the years I've had a lot of failures. It is still a challenging condition for the profession.

Building The Clinic

As the years marched slowly forward and I was called upon more and more I began to realize the developing challenge of an area sparsely settled and widely scattered. Travel time from one call to another increased the costs to me and the client. The big problem was my time. I could see the need for development of good facilities to handle the large animals and then encouraging the ranchers to bring in their sick and problem cases.

I didn't have any money accumulated, but when I talked to the banker my credit was still good. I felt I needed to be on the highway, and the ideal location was someplace between town and the sales barn. I first negotiated with the city to put some of their property next to the state road shed up for sale. They agreed, but when the bids were opened, Leland Stevenson got the land so I had to look elsewhere. The Wintertons owned the land north of the highway just past the cemetery and I was able to purchase thirty-four acres from them.

Dad and I sold the cows with the permit, and an eighty-acre piece of our farm. We paid off the mortgage on the cattle and had enough to pay for the land. Now the problem was water. There was an old well back by the hill, which I questioned would furnish enough water. The other option was hooking onto the city system at the state road shed. I debated with myself and negotiated with the city until they finally agreed to let me hook on, if I would put in the line at my expense, give it to them, then pay one and one-half times the usual rate.

The running of the line around that hill became a real challenge. We ran into ledge rock before we were past the road shed property and it continued to within about one hundred feet of my property. I was fortunate to have my father's expertise with dynamite. I rented a compressor and ran the jackhammer myself to drill the holes for the dynamite. We would load the shots, then stop the traffic and blast away, then hurry and clean the road up. Ron Phillips did the digging with his big backhoe. Sometimes we had to shoot the rock three times to break through all the layers. I became quite adept as a jackhammer man and even did some of the blasting when Dad wasn't there, but it was a costly and challenging way to put in a water line. I'd clean up after a blast, take off on a call while Ron dug things out, then come back and go at it again. When we finally got past the ledge rock I took Ron and showed him where I wanted the pipe to end and left on another call. When I got back I found he had dug the trench on an angle into my property instead of straight in and this created a problem in locating my clinic. It required the clinic to be closer to the property line than I wanted, but I thought that was better than building over an unsettled trench that might crack the foundation and floors.

When it came time to lay out my corrals and shed I didn't have any reference to arrive at somewhat true directions. I didn't want to hire an engineer. It was the 21st of September and I remembered that the sun should set directly over the equator on that day. It would give me a true east-west line if I used the sun and a shadow. That evening as the sun was setting over the hill to the west I took a long 2x4 and stood it at the far end, then sighted in on the shadow and the sun and drove some pegs. It may not have been exactly east and west, but it was close enough for me.

Dad and I went out and cut some large cedar posts and we hauled poles back from the farm. When we purchased the cattle I had gone up on Farm Creek Mountain and cut several hundred poles to build corrals on the farm. There were a lot of them left over, so we used them to build a good set of corrals along with a shed. I had a good place to unload the critters and a chute to handle them. Mom, Dad and I made a trip to my brother's place in Iowa, then I took the truck and went on back to Indiana to pick up my squeeze chute.

When I got ready to start building the clinic, I engaged my cousin, Ted Thomas Jr. His help and expertise saved me a lot of money. I did a lot of the work myself under his direction, which saved me a lot of labor costs. Elvyn Bascom laid the block, Junior Hicks helped with the wiring and Lawrence Pike did the plumbing.

Dr. Bill Arters came to work with me to do his internship and get his license that summer. His two brothers came out from Ohio to visit him during the winter. They were expert sheet rock men and while they were here they did the ceiling.

As I look back upon all the effort it took to put in the water line, cut the posts and poles, dig all the post holes, build the corral, shed and clinic I don't know where the time and energy came from. I couldn't have done it without my father's untiring help, and the help from others. I was really blessed, for within a year, I had the facilities I needed. The monthly payment on the bank loan kept us scratching for many years, but it proved to be a wise investment.

7
GETTING THE BABIES HERE

Dystocia

Dystocia, or difficult birth problems, always sent a wave of apprehension through me when I received a call for help. The question of what the problem might be and the challenge of being able to correct it, coupled with the urgency involved, placed a burden upon me which I couldn't take lightly. Because of the nature of my practice it most often involved cows, but mares foaling came along, as well as ewes with lambs, and pigs in trouble.

The management and correction of these problems were truly an evolutionary process, both for me and for the owners. I learned from doing and they learned to call on me early enough that I could save the lives involved.

It seemed like the standard procedure was to reach in and hook onto anything they could find with a few strands of bailing twine, then tie the cow's head and hook on to the calf with the truck or tractor; or try a little bit, then wait until the calf was rotten and stinking to call me. In fact, if it weren't for these odd requests, many of them would probably never have learned that I could help them at all. Word of my competence was slow getting around and very discouraging for many years. I think their own pride made it hard for them to acknowledge that I might know more than they did. Correspondingly, it took them some time to learn what they could do and when to seek help.

Several years after we moved into our home I realized that I needed a place out of the cold where I could do the operations. I'd been pulling calves and doing C-sections in the trucks out in the back lot where all the neighborhood kids got an education, but when it was cold or storming, both the cow and I needed a shelter. I talked to my dad and he and Wes Jensen came over and built a frame garage with just siding and tar paper on the outside. It had a dirt floor and some old timbers from the farm for a foundation. Before winter, I insulated it and lined the inside with masonite. Then I bought an old oil stove, had a low table given to me and moved the dog kennels out there. With a couple of posts and some plank, I made a small unloading chute at the walk-in door. It was warm and I could park my truck in there and not have to take out the medicine that would freeze. It didn't look like much on the outside, but it had good lights, power for clippers, and was painted all white on the inside. I felt I was right uptown. In the years that followed I did quite a lot of C-sections in there.

My neighbor across the alley, Dee Allred, used to come over and assist me. One night a fellow brought in a little heifer that needed a C-section. We pulled her out of the truck and rolled her up on the table and tied her feet. I clipped the hair off, scrubbed her up, disinfected the area and proceeded to deaden her. I noticed that the owner turned away and didn't watch much. We needed his help to pull the calf out of the incision, so he came over and helped. As they took the calf out and laid it on some sacks, Dee said, "You better lick it off," meaning rub it down with those sacks and get it dry. I don't know whether the guy took him literally, or whether it was just the blood and smells, but he grabbed his mouth and bolted for the door. He was sick and turning green.

Client Feelings

Working alone created some problems in client relationships when I would have to change my priorities from the more routine work to cases of more urgency. Most clients were very understanding and many volunteered or went along to help if they could. Dystocia calls came most often during the evening and very early morning hours, or at 10:00 p.m., midnight and 2:00 a.m. when alarms were set by the owners to check their herds.

I always tried to ask the owner when he called what his assessment of the problem was, and whether it was an older animal or a first time birth. Other questions I asked were if they could feel or see two feet or a head started, and was the baby alive or dead and beginning to smell. It was surprising to me to hear the varied answers. Some owners didn't really know much about what their problems were and others knew exactly what was wrong.

In the beginning I was always expected to come out to their place even though we would often have to load up the heifer and bring her into the garage if we had to operate. Later on, especially after the trailers came into being, I would get a call to see if they could bring in their animals for help. I'm sure, having facilities, together with a few past experiences, encouraged owners to come to me.

The temperament of the patient varied from the downers that wouldn't try, to the ones that were wilder than deer. They would fight and buck until squeezed in a chute, or thrown down with ropes. There were mean ones that wanted to eat you alive, and relatively gentle ones that would wait until you took hold of the tail, then kick you hard enough to nearly break a leg. The most exasperating aspect for all types, was the way they strained and pushed as you were trying to chain up or manipulate a leg or head in preparation for delivery. Just as you reached in the full length of your arm they would bear down and push, squirting urine, manure and uterine fluids up your sleeve and down inside your clothes. Sometimes they would catch your arm or hand between the edge of the pelvis bone and the baby's head. The pressure they could generate would stand you right on your tip toes with pain. Just when you were about to turn a head or straighten a leg, the push would come and make it impossible. Then when you got things all lined up and began to pull they wouldn't help you at all. There are few, if any, things I was called upon to do that required more hard work, patience, strength and stamina than delivering calves or colts. Hand grip and finger dexterity were taxed to the very limit and arms become effectively paralyzed in prolonged efforts. So many times I wished I had two hands on the end of my arm, or wished my joints would bend the other way.

When I first started out, each cow was a new mystery. I had to find what was preventing the calf from being born and then work to correct it. As time went on, I began to realize that patterns were emerging and many conditions repeated themselves. As I examined a cow I quickly learned what to check for and what to expect. Were the limbs I could feel back legs or front? Was it alive? Did minor abrasions and swelling tell me the cow had been worked on extensively? Was the calf in a normal position, or coming backwards in a breech position? Was the head turned back and how big did the calf seem in relation to the pelvic opening? Probably the most frequently encountered problem was first-calf heifers with a calf too big to be born. Most of the other problems fell into about four malpositions, the most common being one or more front legs back or up over the head. Next were heads turned and rotated upside down, then calves coming backwards, and finally the tail first breech presentation. There were other things like twins and triplets mixed up with each other, and a few deformities like two headed calves, siamese twins, fluid distended abdomens and a condition called shistasomas-reflexus where the head and all four feet point the same direction from a very shortened spinal column. Occasionally a calf would get turned over, uterus and all, causing a torsion or twist in the birth canal.

The first-time heifer with a large calf to deliver presented my clients with the most challenges and they were approached in many different ways. The solution ranged from a bullet between the eyes, a chain from a tree around the neck and a tractor on the calf, to an embryotomy, or a pelvic symphysis split with a chisel. With patience and education I was able to convince most that a C-section done in time gave them a live calf and a cow in good condition to care for it. In a few cases where the calf was already dead and especially if it was getting putrefied I would cut off a head, or a leg, and make it possible to pull. The procedure I used most often on a dead one was to reach as far up on the shoulder as I could with a small hook-bladed knife and cut the skin down to the foot and around the leg. I would then loosen the skin and pull out the leg with muscle and bone intact. With that much reduction in size the calf would usually pass through the birth canal with the help of a calf puller.

The pivotal question that soon emerged in delivering these first-calf heifers was whether the calf could be pulled through the birth canal and survive, or whether it should be taken by C-section. After a year or two of trying to make that decision I finally developed a criteria that was correct for me in nearly all cases. I would put a chain around the head behind the calfs' ears and then down into its mouth. With that accomplished and the feet in normal position, I would have the owner, or myself pull on the head. If the head would slip into the pelvis with the pull from just one man, the calf could be pulled alive. Otherwise I would do a C-section. If the calf was already dead, more pressure could be applied, but there was always the danger of damage or paralysis to the heifer, especially if it was an extended period of time.

I preferred to take the calves out through a midline incision, from the naval to the udder, on the beef heifers. The abdominal lining is thin and strong there and it seemed faster for me. I would do them on the side if the owner preferred and let me know. On the dairy animals I always used the left side approach to avoid having an incision and tenderness near the udder. It really didn't matter much to me which approach I used—they all healed up in a week or so.

I remember one night I got a call from the McConkie's up at Altamont. They had a heifer they couldn't get the calf from. It was late at night and when I arrived they had the heifer tied down in a small pen with a shed on one side. I had an idea that considerable pressure had already been applied to the calf, because they said it was dead. I could also see that their calf puller had a severe bend in it. I lubricated things up good, put the chain on the head and hooked up my puller. I could tell very shortly that the calf could never be pulled and I only tried because we didn't have anything but a fence and a gasoline lantern to do a caesarean by. She was wild and fought our moving her over to the fence to get her legs stretched out. I clipped a little of the hair off with my scissors the best I could, then washed her up and applied the antiseptic to the area. It didn't take me long to deaden the area and cut out the calf, but it was a bit more difficult to suture up the incisions with only the lantern to see by. When I got through my legs were half asleep from kneeling down there beside her. I went to my truck and got a shot of combiotic that I gave to her, along with instructions to repeat the shot each day for the next three days. We gathered up all of the equipment and took it out of the pen, then took off all the ropes and let her go. She rolled over on her tummy jumped to her feet and made one quick circle around the little pen before she sailed over the fence and disappeared into the night. She didn't get any more antibiotic shots, in fact they didn't find her for ten days, but they told me she was doing fine. That taught me that surgery could be performed under rather primitive circumstances and the patients still survive and over the years there have been some strange operating rooms.

Another night, (it seemed like they always came at night) I was called about a heifer that couldn't calve. The owner had her in out of the night in a small milking barn. They said they tried for a short time but couldn't get the calf so they called me. When I washed the heifer off and felt inside it was dirty and gritty and so swelled up I could only fit my arm through the birth canal. I realized she had been worked on extensively and knew that delivery of even a very small calf would be impossible. I told them a C-section was the only thing we could do. For some reason I had a feeling they agreed almost too quickly and wanted to get it over with. We slid her back up against the stanchions and tied her feet up high, front and back. The floor was concrete, the lighting not too bad and I had power for the clippers. I clipped, scrubbed and disinfected, then deadened the incision area. A few bold strokes with the sharp scalpel and I had the calf by the hind feet. When I pulled it out and laid it over on the floor, I noticed the front part of the bottom jaw was missing. It had been pulled off leaving the two jagged edges of the jaw bone sticking out. I didn't say a word, but we did exchange glances and it was very quiet as I sutured up the uterus and abdominal wall. By the time I finished with the cow the calf was standing up. After they wrote the check and I was ready to leave, I pointed to the calf and said, "You guys will have to finish the job with him."

Two Headed Calf

Cesarean sections were performed primarily on heifers, but occasionally some conditions encountered in the older cows required this approach. One I remember well was a cow brought in by Rue Miles with a calf that had two heads. It was a normal calf up to and including just two ears, but from there on forward everything was duplicated. There were four eyes, two noses, two mouths and two tongues. The skull was fused together forming about a sixty-degree angle between the faces. When I felt those two faces spread as far apart as they were, I knew it couldn't pass through the birth canal. I did a C-section and took the calf alive. It was quite a novelty and generated a lot of attention. People far and wide came to see it and school classes had field trips to bring the children. The calf could never stand, it had too much head to hold up, and it couldn't eat by itself. The tongues had a common root, but forked just forward of the larynx, causing the calf to choke on milk put in either mouth. I fed it by stomach tube to keep it alive, but the two inside eyes lacked normal lacrimal glands to keep them lubricated. This resulted in some pain and corneal opacities in about ten days. I sent the calf up to the veterinary diagnostic laboratory where it was euthanized and studied. They digested all the soft tissue off of the bone and sent me back the fused skull, which I still have. It has been the subject of show and tell many times over the years.

Uterine Torsion

The other condition that frequently required the knife approach was torsion of the uterus. Whether the cows got knocked over in the herd, or began frolicking around bucking and playing, or for some other reason I wasn't able to visualize, they succeeded in flipping the gravid uterus over, causing a twist in the vagina. Characteristically these cows never showed much active labor and seldom got down to pushing. They wandered around with their tail out and didn't show many signs of calving. The owners wondered what was going on and usually called for help. When I would insert my hand into the vagina, I could feel a rifling effect of the twisted walls as I went forward. Sometimes it was possible to reach clear in and get hold of the calf, but it would be impossible to deliver it. On one or two occasions I was able to grasp the calf and move it in a circular direction and correct the twist. Most of them were too heavy to turn. I even tried rolling the cows over when sufficient help was available, and also rolling them on their back with my arm inside to try and turn it. I don't recall any successful attempts with that approach.

One approach that I did use was to shave and prepare the

cow for surgery on the left side, then make an opening large enough for my arm. I could reach in through the opening and rotate the uterus back to its normal position. Once this was accomplished, I would temporarily close the skin with towel clamps and deliver the calf normally through the birth canal. This was always a little scary because the cow would bear down trying to push the calf out and force her digestive organs up against my temporary skin barrier. The pressure they can generate is hard to contain. I was always between a rock and a hard place, because I didn't want to close up the opening until I knew the calf could be delivered, and I didn't want her tearing out the stitches with the extreme pressure of laboring. Many of them I would go ahead and do a C-section on because I felt like the calf had the best chance of survival that route.

Breech

Another condition where the cow may not show much labor is when the calf is coming tail first in the true breech position. The back legs are up along the side of the calf's body and only the rump is presented at the opening. They can be very hard to straighten out. I learned early that it was necessary to approach it systematically. The first move was to push the calf forward enough to get your hand in. Sometimes this was a herculean task. Next I would work a chain around one of the legs between the body and the hock. Once I had pulled the chain through the ring and made a loop I had to push it down toward the hock as far as I could reach. Now I had someone pull on the chain as I pushed the calf forward. This would pull the leg back some and partially double it, while I held the calf forward. The next move was to get the chain down below the hock joint and repeat the pull while I pushed. This would eventually bring the doubled hock up into the pelvis with enough manipulating. Next I had to move the chain down around the fetlock just above the foot. Then it was critical for me to cup my hand over the hock to prevent a uterine tear while I pushed it forward and a little bit laterally while the foot was being pulled up and out. The hock is narrow and sharp enough to tear the wall of the uterus if it wasn't protected. Once the foot came over the brim of the pelvis and straightened out, I was half done with the manipulating. I repeated the procedure with the other leg and once it was out the calf could be delivered backwards. Nearly all of the calves were dead before I started, but if by chance one was alive it was very important to immediately hold the calf up by its back legs to let the fluid run out of its mouth and throat. I would often take the calf as it cleared the birth canal and drag it up over its mother's body with its head hanging down, then have the owner hold it while I stuck a straw up into the nostril. The tickling in the nostril made the calf cough and clear out its throat. I began doing this when I noticed they would cough sometimes when I was cleaning mucus out of the nose with my finger. Quite often on the old family farms they still have a few chickens providing feathers that were scattered around by the wind. They were excellent to tickle the nose with.

No Head Doc

Another frequently encountered malposition in my experience was a calf with two front legs out but no head. When I would examine them I would find the head rotated to one side or the other and upside down under the protruding legs. I soon learned that most of these were the result of the owners pulling on the front feet before the head was in the pelvis. A heifer with two feet out, and no head to be found, was symbolic of the Chasel place. In spite of the challenges I faced when I went there, I was always impressed with the farmstead and its care. The gates all swung, the fences in good repair and the livestock well fed and cared for. Old Jerry was from Austria and he had brought the ideals of his native land with him. He always impressed me as a quiet, patient and gentle man who accepted life and its challenges without too much concern. Honesty was important to him, and in his dealings with me he went out of his way to be sure I was paid as agreed. He hadn't mastered the English language too well and spoke with mannerisms of his native tongue. I liked him, respected him and liked working for him.

When I was first contacted for help—I think it was a calving problem—he didn't have a phone and had to go up to the Yack family to have them call me. He and his son Frank were running the place when I started going up there.

Now Frank was different. He was a good worker and quick to help, but he was always nervous. It seemed like his spring was always wound too tight. I could tell that he was very insecure and sometimes tried to hide it with bluff and bravado, but at the same time he was seeking approval. He was a loner in his thirties and almost without a social life. He frequented the little beer parlor in Neola and came to town with his dad, but he seldom came alone. He spent most of his time on the ranch, and he had a big old heavy calibre pistol that he wore on his hip much of the time.

It was his nervous disposition that caused most of his calving problems. He was very attentive and watched those heifers like a hawk. When one of them started to show signs of labor or looked like she was going to, he had her in the barn. His main trouble was that as the cow started showing definite labor he would try to help too soon. He would go in and chain up the legs and begin to pull before the uterus had funneled the calf's head into the birth canal, and it would be forced to one side or the other and end up wrapped down around under the legs in an upside down position. He would call me, or bring the heifer down, and tell me the head was upside down. Sometimes in the process of loading and hauling, the calf would slip back down in the uterus and get its head back up by itself.

This was a malposition I encountered many, many times, not just with Frank, but a lot of others. The heifers themselves, even when left alone, would push the head around and under the legs wrong side up. It was a condition that required a specific and methodical approach to get the calf straightened out. I would take the calf's legs, one at a time, and push them back to where I could double them up. Then, carefully so I wouldn't tear the uterus, push the foot down and under the body of the calf as far as I could reach. When I had both legs doubled under it would hold the calf forward and I could take the head in my hand and rotate it back up on top. Once in position I put a chain around behind the ears and in the mouth, then pulled the head into the pelvis if it would come. It was then just a matter of fishing out the legs again and delivering the calf if it wasn't too big. Sometimes owners would tell me how long they had tried to straighten the head out, and I had done it so quickly it was embarrassing to them. I could understand their frustration, because it was almost impossible to get the head up without doubling the feet back under the calf.

While Jerry was still alive they didn't have as many turned as in later years when Frank was alone. I think he could get Frank to wait a little while. He seemed to have a firm control on the decision making. After Jerry passed away, Frank called me a good deal more and seemed to heed my counsel. I could feel his insecurity and tried hard to help him. I thought our relationship was very good until one night Frank called me to see a critter and I already had another call to go on west of Neola. I told him I would stop on my way home. The other call kept me occupied until about midnight and when I stopped at Frank's there were no lights. I thought he might have dozed off while waiting and I thought if the problem was important, he'd want me to wake him. I went to the house, knocked hard on the door and waited for a minute or so. I didn't hear anything and no light came on, so I knocked again. There was no response this time either, so I went on home. He called me the next morning and I went up. He was coming out the door as I pulled up and stopped. He had on his big pistol.

"Did you stop here last night and knock on the door," he said with kind of a chip-on-his-shoulder attitude.

"That was me. I told you I'd stop on my way home."

"It's a good thing you didn't knock one more time. I was on the other side of that door and I had this," patting his gun, "pointed right at you with the hammer cocked. One more knock and I was going to let you have it." I didn't know how to interpret the meaning, and never could come up with an explanation other than he was afraid of the night. I know from that time on I was always a little wary of Frank Chasel, even though I worked for him for years afterward.

Ewes

Because of the commonality of twins and triplets in the ewe, the multiplicity of problems encountered didn't have any limit. I would find many puzzles to work, and twisted bodies to find heads and legs for. The small size of the lambs made them easier to manipulate, but the fragile uterine walls could be torn so easily it was still a challenge. It took a lot of patience and care to maneuver them into position to be born. The legs were unusually long in relation to the body and although they could be manipulated with one or two fingers it took some doing sometimes to straighten them out and not injure the mother.

Another problem that happened frequently was the mother willing to claim only one lamb and refusing to have anything to do with the other one or two. There were many schemes devised to overcome this, from tying the two lambs together to tying the ewe so she couldn't butt or kick. Along with the rejection of their own lambs, it was sometimes desirable to graft orphaned lambs without mothers, or the third lamb of triplets, onto a ewe that could feed two lambs. Like the first case of rejection, many things were devised, but among the most successful was a hefty shot of tranquilizer, accompanied by a liberal spraying of the lamb and the mother's nose and face with a bathroom deodorant. If you could successfully mask out her detection by smell, and keep her doped enough she didn't care for awhile, she would get used to the strange lamb and "pair up" as we called it. Like other facets of life there were always exceptions and no matter what was tried the ewe would never take the lamb.

In the big range herds there wasn't time, pens or help

enough to work with the rejected and orphan lambs, so they became what was called 'bummer' lambs. Many of these were given away, or hauled back to the home ranch to be raised on bottles. There were many farm kids that got the opportunity to raise a bummer lamb, and most, if not all, felt like it was a special privilege.

Pigs

The pigs, like all animals having babies, had their share of troubles. Many times I could reach and pull the baby that wouldn't come with my hand. If it was coming back feet first I could get a finger between the legs, and get hold enough to pull them out. If the head was coming first there wasn't anything to hold on to. I devised a loop of bailing wire that I could push back over the top of the little pig's head, then pinch the two strands of wire close under its mouth with my fingers, while I pulled with my other hand. It was amazing how effective that was. Later on, they came out with some pig forceps that were very helpful. Like all animals there were those that were just too big to be born and had to be taken by C-section. It was a little different where I'd open the long tubular uterus in one place, then have to slide the little pigs along to pop out that hole. With all the dinner spigots lined along the bottom of the belly, I had to use the side approach on the sows.

Because of the relatively low value of the sow and the ewe in relation to the cost of the surgery, even though it was done at half price or less, I didn't do very many operations. With sheep the owner frequently sacrificed the mother's life to get the lambs, and many of the pigs were let go too long to be a good surgical risks. Regardless of how they were delivered, it was always a source of great joy to me to see the miracle of birth and a live, newborn animal immediately begin the struggle to get on its feet and hunt for something to eat. The challenge and anxiety of delivery was rewarded with a feeling of satisfaction and worthwhile accomplishment.

Getting the babies into the world was only part of the problem. Keeping them alive was equally challenging. I came to realize that in spite of all the preparation and planning, the weather could wreak havoc with the newborn animals. Typically the Basin had a hard time warming up in the spring and many calves and lambs were lost from getting chilled and cold. Whenever a new one was born during a cold night and the mother didn't get it licked off and dry, it could suffer from hypothermia. They would be found nearly comatose and it would always be a race against time to get them warmed up and fed. Heat lamps, blowing heaters and stoves were all used, but with the baby still wet the evaporation continued to cool them until they were dry. On some occasions I had owners place them in a tub of warm water to get them warm and fed, then dry them off and keep them warm.

Serious Trouble

When it came to those that were dead and rotten there wasn't much pleasure, because I would get saturated with very putrid fluids and blood, while pulling out the fetus in handsful of hair, rotten meat, and bone. The owners most often stood upwind and as far away as they could. Many of them would get sick and vomit up their supper. The amazing thing was that the cows would live, if we could get the dead calf out.

The most amazing case I ever had was a heifer of Bill Peatross's. He had observed what he thought was the beginning of labor one morning, but she stopped all signs and went to eating. About three weeks later, they saw some discharge coming from her, and when they examined her she smelled bad. When they couldn't feel any calf they brought her in to me. The cervix was nearly closed but I was able to work my hand through. There were a few remnants of the bones, but mostly a putrid liquid that I spooned out of her a handful at a time. That calf had been dead in her for three weeks and she was still alive.

I don't know how my good wife could stand to live with

me, because it was impossible to wash that smell off my skin. I would often use straight Clorox on my hands and arms, but it wouldn't remove all the smell, just replace it with another.

Cold Day To Remember

One cold day with a rather stiff breeze blowing I was content to stay inside and look out at the snow. I was never one to wish my life away, but January was never very pleasant in the Basin. I was anxious for this last week of it to hurry on. When I came back from my lunch I stopped and stood in front of the gas heater blowing warm air. A shiver ran through me as that warm air hit my back. It surely felt good. The phone rang and I walked over to the desk and picked it up.

"Basin Vet Clinic."

"Doc, this is Carnes LaRose. I've got a cow that can't calve, could you come up?" The thought of a cow calving on a day like this in January didn't fit the norm.

"What in the world are you doing calving this time of year?"

"Oh, I'm not calving. It's just one of those things. I guess a bull got in when he wasn't supposed to last spring. I think the calf's dead and she's down and can't get up." He hesitated and waited for me to answer, like he expected me to say, "You better just shoot her."

"Where is she, Carnes?" I wasn't sure where he lived.

"She's in that field east of Leon Dump's place. I could meet you there in the lane that comes by his place."

"Alright, I'll be up as soon as I can get there." I filled two jugs with fairly warm water and took them to my truck and set them in the cab by the heater. When I went back inside and felt the contrast in temperature I decided to fill two more jugs with the hottest water I had. I pulled my overshoes on over my boots and buckled them. I had my coveralls on with the short sleeves, so I slipped on my lined Levi jacket and headed out.

The road was bare all the way to the Ravola dugway and I clipped right along, but from there on it was snow packed and

I had to go a little slower. There were little sheets of snow starting to drift across the road in spots when the wind gusted. I thought to myself, "I hope that old cow found some willows or something to break the wind." I took the road past the creamery and found the lane. It had been plowed out but the snow was drifting behind every lump. It was a good thing the wind was blowing straight down the road. I spotted Carnes's pickup parked at the far west end where there were some big trees. I think it was a farmstead at one time, but the house had long since burned or been moved and a few old posts and the trees were all that remained.

I looked around for a cow as I pulled up next to his truck, but I couldn't see one. I got out of the truck and took out my little calving kit. It was a small aluminum suitcase in which I kept my chains, the head snare, a bottle of homemade lubricant, some hooks, my hook-bladed knife and other tools for delivering obstinate baby animals. Then I reached in the back and pulled out the calf puller which was in three pieces. Carnes stepped up and I handed him the kit and the two long bars. I threw the britchen chain over my shoulder, tipped my hat into the wind and moved up to the passenger door and took one warm and one hot jug of water out. It had cooled off some. We had everything, so I shot a questioning look at Carnes and he motioned toward a small open gate.

"She's down in the field there." He pointed with the puller rods as he started wading through the snow toward the gate. About a hundred and fifty to two hundred yards out toward the center of the field I could barely make out a little dark spot in the drifting snow. I thought maybe we ought to have a rope. Sometimes those cows that were down and couldn't get up had led me on some merry chases. I went back and got my lariat and followed him through the gate. The snow was about fifteen inches deep and crusted some on the top making our progress slow. When we approached the old cow she was laying with her head downwind and the snow was sifting up over her back and forming a small drift in front of her. She was almost white with the drifted snow clinging to her long hair.

It was no wonder I couldn't see her. She didn't make any effort to move and the look in her eye was a little dull.

"I guess I shoulda just shot her," Carnes said a little apologetic, with a question in his eyes. I guess he wondered if that still wouldn't be the best thing to do under the circumstances.

"I think we better give it a try, we've come this far. You never can tell whether they're going to make it or not," I said. My never-give-up attitude welled up to defy the wind and cold.

"Yeah, I guess, but she probably won't make it here, if we do get the calf out of her," he said with some apprehension.

"That's for sure. You may have to get a tractor and a slip or something to take her in where you can feed and take care of her," I said as I kicked the snow away from the back end of her. We cleared a small area and put down the things we had in our hands. I put my hands in my pockets to warm them a little, then peeled off my coat and dropped it on the snow we'd kicked back. The wind hit those bare arms like cold water and I wished that I had put my sheepskin vest on under my coveralls. I knelt down behind her and began washing her off as fast as I could. As I leaned forward those blowing snow crystals pelted the back of my neck like little needle pricks.

As I felt inside I found a tail and that meant a breech presentation and the butt of that calf was pushed into the pelvis as tight as a cork in a bottle. I knew I wouldn't have much chance of pushing it forward unless we could get the cow's legs stretched out behind her and turn her up on her sternum.

"Where's the rope?" The snow had already partially covered it. "Let's roll her over and get the rope on her back legs," I said as I started pushing her over. The old gal was either cooperating or so far gone she didn't care, because she laid over on her side in the snow. We got the rope on the two hind legs and I pulled them out behind her.

"Now, get her by the tail and roll her up again while I hold these legs," I said. He had to try a couple of times, but again she helped by throwing her head over and straightening up.

"Now you hold these legs while I push that calf back in." I handed him the rope and I flopped down behind her. The smell was getting up to my nostrils now and then, telling me the calf had been dead awhile. I started to push and my feet slipped in the snow as I arched up to put the pressure on.

"Stand on my feet while I push." I indicated my problem. When he was braced on my boots I gave a heave and the calf slipped forward. I could get my arm in now. I uncovered the kit and pulled out the lubricant, a long chain, and handed a D handle to him. As I pushed my arm in up to my shoulder trying to feel around a leg, the usual began to happen. The fecal material heretofore blocked off, could now flow freely with the calf out of the birth canal, and the normal reaction of the cow was to push and strain. It gushed out over my shoulder as I was feeling around the upper part of the calf's leg. After a quick rinse with the now cool water, I took the chain end in my hand and entered again. It was a tough challenge to work the end of that chain around the leg because the uterus was so tight around that calf. I finally made it and put the other end through the loop and tightened it on the leg as far forward as I could reach.

"Now I'm going to push on the calf's butt and I want you to pull on the leg at the same time." He hooked the D-handle into the end of the chain and braced his foot. I put my right hand in and pushed as hard as I could while holding onto the chain with my left hand. When I had the calf as far forward and somewhat to one side I nodded my head for him to pull. The hock of the calf came up over the pelvis and popped back into the birth canal.

"I think we about got her made," I said with a grin, because once you get one leg in that position it holds the calf forward. "I just gotta get that other leg up and I think we can pull 'im." The calf was held forward, but now there wasn't much room for my arms, I discovered after rinsing off the manure and diving in again. I tried pushing the end of the chain around the leg from the inside out against the uterine wall. It was tough. I tried several times, but I couldn't put enough pressure on the chain loop with the ends of my fingers. Finally I gave up, pulled the chain out and washed up again. The wind hadn't slowed up a bit and the crystals pelting my neck were still sticking me like needles, but I was working so hard I hadn't noticed. On my next try I worked the chain around the leg from the other direction. I could barely feel the end of it as I checked the inside, but by giving a little extra push I hooked one finger in the loop and was able to pull some slack. I made the loop and was able to work it down over the point of the hock. With a push on the calf's pelvis and Carnes' pull on the leg, it came up into the birth canal. I heaved a sigh of relief and would have taken a moment's rest, but that wind put me right back in gear. Now I had to work the chain loop down over the fetlock just above the hoof. It was a methodical approach, I'd learned how it had to be done one step at a time. When I had the chain in place on the foot, I then had to push the hock forward far enough to let the foot slip over the rim of the pelvis. This was the ticklish part, to push it in against the pelvis of the calf so the point of the hock didn't tear through the wall of the uterus stretched so tight around the swollen calf. The first leg came out with a jerk as it cleared the rim of the cows pelvis. "Only one more to go," I thought as I checked the uterine wall for a possible tear. The room in the birth canal was limited again, but I got the chain on the leg and pushed it down to the foot. This was the tough one, because the other leg was partly in the way. I took hold of the leg as near the hock as I could and twisted it sideways as I pushed it forward. When I thought it was about far enough I indicated for him to pull. The chain came loose and out of the cow bringing one of the calf's hooves with it. I grabbed the other one just inside and pulled it out too. "That calf must be pretty rotten," I thought to myself. Anyway on the next try we got the leg out and I looped both legs with the chain ready for pulling.

"Okay, if you can find the puller, I think we can get him out." I scanned the drifted snow where the puller and kit had been. He kicked around in the snow and found the puller. We decided we'd better get the cow's back feet in a normal position before we pulled. I pulled on her tail and he pushed her hips and rolled her onto her side again. That was a good position to pull the calf, so I hooked up the chain and took out the slack. I had to dig around in the snow to find the bottle of lubricant. That snow was drifting faster than we could kick it out. When I had worked a big handful of lubricant around the calf as far as I could reach, I started to pull and the old cow began to strain as well. Things moved right along until the hips were about through the pelvis, then things began to tighten up. That calf wouldn't budge. Because of the rotten condition of the calf, I was afraid it might pull apart, so I released the tension on the chains.

"I'm afraid of pulling the calf in two, I think I'll try and get some more lubricant around it," I said in explanation.

"Maybe help." Carnes was part Indian and didn't talk very much. I could tell he was getting cold just standing and watching me work. When I pushed my hand up between the legs with a hand full of lubricant I could tell what was wrong. The abdomen was swelled up like a balloon with the putrefactive gasses. Being in up to my shoulder with my face down close, there seemed to be a strong odor of gas coming out of her around my arm. It wasn't hard for me to turn my head and get a breath of clean air; it was plenty fresh that day. As I reached up toward the umbilicus I could feel a vibration and realized that gas was escaping from the calf's abdomen. I grasped the placental membranes that were in the area of my hand and pulled them out hoping I might break the umbilical cord. I thought it helped, because the gas seemed to come out a little faster, but this was not a time for patience.

"The calf's belly is all bloated up, that's why we can't pull it, but it's leaking out through the naval. We'll just have to give it a little time," I said. The skin on my arms felt like it might be freezing. I decided I'd better get a coat on. "Grab that other jug and wash me off." It was the hot water to start with but was now just a little cool. By the time I got the worst of the muck off those bare arms were freezing. I found my coat in the snow and shook it out and pulled it on my wet arms. What a relief to the arms but it made me aware that my feet were mighty cold as well. I jumped up and down and banged my arms around me for a few minutes, then dropped down by the puller and began to work the jack. It got pretty tight, but I was to the point where I felt like it was going to have to come out, or we would have to shoot the old gal. I didn't think she would survive a C-section and it looked about impossible to get anything in there to load her anyway. I added a little more tension and pulled down on the puller. Out came the calf with a gush of fluid, placenta, and a stream of fecal material. The old cow had added her push at the right time and out it came.

When it happened the cow rolled up onto her sternum with her feet under her. We began to kick around in the snow to find the things we'd brought and I took the chain off the calf's feet and dropped it in my pocket. Carnes handed me the D-handle and the lube bottle as I opened the kit and then closed it. The rope was all strung out in the snow and when I began to shake it and coil it up the old cow looked back at me and slowly got to her feet. She was a little weak, but she started off down through the snow toward the other cows. The temperature was getting colder as the afternoon approached evening. Our tracks were all drifted full but we gathered up our stuff and slogged back toward the trucks, bucking the wind as we went. We put the things in the back of my truck and turned our faces away from the wind.

"Thanks, Doc. Be alright if I stop in and pay you?"

"That'll be fine." I was so cold getting paid didn't matter that much. We both just wanted to get in out of that wind. It was a few miles before the heater in the truck began to work, and after it did, I couldn't stop chilling. My feet were like a chunk of ice.

The Spirit Of Helpfulness

The spirit of helpfulness between one or more neighbors helping each other ran very high in this area and may have contributed to the problem of delay in calling for my help, but as the years progressed and the understanding of my capabilities increased it became a source of great help to me. I have come to cherish and appreciate the many wonderful friendships that evolved as a result of their helping me with my work. It has become life's most satisfying facet to feel the friendship and respect of everyone that I meet. It's a great feeling and I wanted to point out how neighbors were cared for by one somewhat amusing incident.

It was a Sunday morning and my wife called me at the church to tell me that Clair Winterton had a mare foaling and needed my help. The message of a mare having trouble always caused me great concern, because their laboring is more violent and sudden than other animals. I rushed home, changed my clothes and got on my way. When I arrived, Clair and I drove out into the middle of his big field where the mare was. It was a beautiful day in the spring, when the temperature had warmed up to where you didn't need a coat and yet it wasn't hot. Clair explained that it was his neighbor's horse and he didn't know how she got in his field, but he found her there and could see she was in trouble. He'd tried to call the neighbor and found they were out of town.

As we approached the mare she stood up and I could see a considerable swelling of the posterior abdomen. She was a good old gentle mare and I soon was able to tell that she was herniated. The pre-pubic tendon was torn all across one side, and the colt was partly down into the hole. I knew it wouldn't be good for her to push and strain, but if I could pull the colt out it would help her, so I washed her off, lubricated my arm good and felt inside. The only thing I could feel was the colt's ribs and back. The feet and head were completely out of reach, and it felt like it was coming in size extra large. There was only one thing to do, and that was to do a C-section, and maybe in the process I might be able to fix the hernia as well. I had never attempted that operation on a mare before, or even seen one, but it was the only choice. There we were out in the middle of the field with nothing to tie her feet to but my truck, and I had only Clair to help me, when I could have used at least two more good men. We decided we'd just have to do the best we could, so I hooked up an IV and started giving her a general anesthetic. I thought the colt was already dead, so it wouldn't matter to it.

The old gal started to get a little wobbly, then just folded up her legs and rolled over like you'd asked her to lay down. I gave Clair the bottle and dashed over to the truck for my surgery kit. When I got back, the nystagmus in her eye was slowing, so I clamped the tube and went to work. She didn't have much hair on her belly, so I just scrubbed and disinfected the area and was ready to cut. Clair pushed her feet back a little and stood against them ready to help me get the colt out. I made a bold cut and kept going right on into the uterus not worrying about the hemorrhage which didn't amount to be much anyway. I got the colt's feet and head started and Clair pulled it out and dragged it over a little ways on the grass, and it was alive. He wiped its nose off with his hand and came back to help me. I pulled the placenta loose from the inside of the uterus and it came quite easily as the uterus contracted, which surprised me. Clair helped me hold things and I hurriedly sutured up the uterus and tied the end of the suture good and slipped the uterus back inside the incision. There were big globs of clotted blood where the hernia had torn, and I removed most of that and felt along the front of the pelvis for something to suture the wall of the abdomenit back to. There just wasn't anything I thought would hold the suture enough to close up the tear. I was at the end of my rope and didn't know where to turn. I asked Clair to lift up her top back leg to see if I could feel a little better, and when he did, that old sister kicked with both feet so fast he couldn't get out of the way. She caught him right on the side of the knee. I lunged forward up against her and got to my feet to give her more anesthetic. Clair was doubled over in severe pain as I grabbed the anesthetic bottle to start it running and asked him if he thought his leg was broken. He shook his head as the old mare took a deep breath and relaxed. It was her last spasm that got Clair on the knee and when she didn't breathe I assumed she was dying, and pushed on her chest wall with my foot. I said to Clair, "I think she has solved my problem, I don't have to sew her up." He was hobbling around, testing out his knee and he said. "Anyway we got a live colt."

The colt had just made it to its feet and as we both looked over at it, bubbles were coming out of its nose. In another minute or two it toppled over dead. We looked at the old mare with her guts hanging out, the colt dead, with its nose full of bubbles and I guess it was so pathetic it was funny. We started to laugh and Clair said, "Well, you can't say we didn't try."

8
PROBLEMS OF THE AREA

Actinobacillosis—Lump Jaw

The Basin is unique in many ways, like the Uintah Mountains running East and West, flat benches at varying levels etc., but it also has a much higher incidence of a couple of animal problems than any other place I've heard of.

Actinobacillosis, or lump jaw, was something I began to see a lot of when I started practicing in the Basin, along with a forage poisoning causing brain damage. They presented a challenge early as I tried to find answers and solutions for the cattlemen.

The lump jaw problem was exemplified by one farm I learned of from the neighbors who saw it as a threat to them. The man let the problem go until he had thirty head with it. The thinking among livestock men at the time was that it was caused by the beards or awns of a grass that grew wild in the pastures commonly known as fox tail. They thought the awns worked their way in through the mouth and caused the lumps, but the problem is really caused by a persistent but slow-growing organism that infects the lymph nodes around the face and head causing abscesses. Because of its slow development the body lays down a lot of scar tissue around the infection and it grows into huge lumps filled with thick yellow pus. The usual treatment was to lance open the abscesses and drain them out and treat the animal in the vein with sodium iodide. The problem was with all that scar tissue, the lumps remained to disfigure the animal. The early treatment had been feeding of potassium iodide in a powdered form before the advent of sodium iodide solution. I reasoned that if iodine was the treatment of choice then iodine deficiency in the diet must contribute to the animals' susceptibility. I therefore began to recommend iodine supplements to the diet for this area, especially those herds that were having problems, and this did help reduce the incidence. As I battled the problem over the years I became convinced that keeping infected animals in the herd increased the incidence as well. I therefore recommended isolating the infected animals or selling them, neither of which seemed easy. They didn't have anyplace to keep them separate, and they clung to the hope that it would clear up and the lump go away so they could get more money for them. It seems like we are all reluctant to take the lumps and bumps of life until they are forced upon us.

Another location for the growth of the same organism was inside the tongue. This caused it to swell up several times its normal size and get very hard, which earned it the name of wooden tongue. The tongue is in fact so hard it feels as if it were wood and the animal soon cannot eat and sometimes cannot drink. Prompt treatment is necessary to save the life of the animal. In handling these cases I learned that although one treatment helped the animal, it didn't heal them and it took two and sometimes three treatments to keep it from coming back.

I could turn my memory back to when I was growing up next to the Nutter Ranch over at Myton where I watched a lot of cows die of this malady because they were not treated.

I had another lesson to learn about this organism when it infected the posterior nares of the nose. The animals would show extreme difficulty breathing and the sides of the mouth would balloon out with air two or three times before they would open their mouth to breathe. After several breaths they would close it and try to move the air through the nose again. I was amazed how long and hard they would resist breathing through the mouth. With this struggle going on they didn't have much time to eat and went downhill fast. I had learned that when putting your hand in a cow's mouth it was imperative to keep your fingers from getting between the grinding cheek teeth or they might get bit off. This lesson came from cleaning wads of feed or foreign matter out of the throats of some animals. Since it was difficult to see the problem in these mouth-breathing cows, I decided I would try and feel back in there. I found that if I could keep my hand pretty much in a vertical position they couldn't bite my fingers, but they could bear down fairly hard on my arm with the bottom front teeth and dental pad on top where there aren't any teeth. Sometimes the pressure would make me squirm a little, but I could stand it long enough to feel the posterior openings from the nose. If the Actinobacillus infection was in that area, those openings were swelled shut and you couldn't get your fingers up into them.

I found the intravenous iodine very effective on these animals, with relief coming in about twenty-four hours. The only problem was an inevitable relapse sometimes as much as nine months later. I would treat them again with positive results sometimes, but eventually the medicine didn't work any longer. It was a hard lesson to learn and one that usually took a second treatment to convince the owner, but I learned that some three weeks to two months after the first treatment it was a good idea to get the animal slaughtered.

As the years progressed I began to see another form of

lump jaw as the result of lancing the abscesses. Instead of the yellow thick pus of the Actinobacillus organism I would find a lump filled with a more liquid, whiter colored pus with a strong necrotic odor. They were usually under some pressure and would squirt out through the hole and get all over me if I didn't take adequate evasive action. It always got on my hand and pocket knife and was difficult to wash off sufficiently to remove the smell. I didn't peel any apples with my pocket knife for a few days after one of them.

I attributed these so called soft abscesses to the Pseudomonas organism and found they responded very well to antibiotic treatment. They would go away without leaving any swelling and little scar tissue.

Photosensitization

"Hello, this is the Basin Veterinary Clinic."

"Doc, this is Hun, up here in Mt. Emmons. I got some sick steers. I ain't never seen anything like it."

"What are they showing you? Tell me how they're acting."

"They act like their legs are stiff and they don't want to move, and their skins cracking and curling up over the hips."

"What color are they?"

"They're Holstein steers about a year old. What do you think it is Doc?"

"Well, I think I know what it is. Did you notice where that skin was cracking? Was it where the white and black skin came together?"

"Yeah! It was, now that I think about it. It was the white skin curling up. What the hell is it Doc?"

"It's a condition called Photosensitization. It's caused by the animals eating certain plants that contain a substance called chlorophyllin. This is metabolized and absorbed into the blood stream and circulated throughout the body. When it passes through the white skinned areas the radiation from the sun affects it, causing a blistering of the skin. It's kinda like a sunburn only it's deeper and much more severe. Where there is pigmentation or black coloring to the skin the color shields the sun rays out and it doesn't happen there."

"What can I do to treat them for it?"

"There isn't much in the way of treatment, especially at this stage of the disease. If a person notices it when the skin is just red and thickened in the early stages you can help by getting them out of the sun, but if the skin is cracking the damage is already done and it won't do much good."

"You say it's a poisonous plant they've et? What's the plant? I've just had them in this little pasture here by my house."

"It isn't just one plant. There are at least twenty-three known plants that cause it and probably a lot more we don't know about. Some of our pasture plants like Alsike Clover can cause it. One thing that makes it bad in our area is the altitude and the clean air which allows the sun's rays to hit us with so much more intensity."

"It's just like a sunburn then?"

"Yes, except it's a lot worse because it happens deeper in and under the skin."

"Would it do any good to lock them in the shed out of the sun?"

"I'm not sure it's going to do any good at this stage of the game, but it sure won't hurt anything. You can put them in during the day and let them out at night to eat."

"I don't think they'll go out to do any eating. I got them into the corral but they don't want to move at all."

"If you go feel their legs you'll understand why. I'll bet the white skin on their legs is dry and hard like a rawhide boot. They don't want to move because it means bending that area of swelling and tenderness."

"What can I do for the poor buggers? Is there something you got that would help them?"

"Not much. Do you have any bag balm or ointment? About all you can do to help is put some ointment on the blistered areas to help keep them soft while they heal up."

"I got a little bag balm and some Watkins salve. Will that do?"

"Yeah, they're all right. I've got some stuff called Moruguent ointment that might work a little better if you get down this way."

"I'm sure I'll need some. From the looks of them critters I'll need a tub full."

"If I get a call up that way I'll drop you off some and take a look at them."

"If you think you could help them, I'd sure have you come up now?"

"Well I don't think I can do any more than you can."

"Okay, Doc. Thanks a million."

He came and got some of the ointment and did his best to treat them. I stopped in a couple of times in the next few weeks and it was a pathetic sight. There were two steers down with half or more of their skin peeled off or peeling and the tissue underneath was infected and oozing serum and blood. They refused to eat or drink much and he finally ended up killing them. On the other steers there were patches of skin that came off and where it didn't peel you could feel a ridge where the black skin met the white. The white skin was about twice as thick as the black and felt hard and stiff.

This was a condition I saw many times over the years, although most were not as severe as Hun's had been. It manifest itself in most of the white skinned animals of the area to some degree. Mostly it was perceptible only as a slightly thickened white skinned area next to red or black skin that was unaffected. In the white-faced cattle with white on the feet and legs, their heads would peel and the legs get sore and cracked. One of the most devastating aspects of it was the blistering of the udders and teats, causing them to turn a deep reddish brown and become so sore they wouldn't let the calves nurse for several days. In the spring when the cows were crazy for anything green after all winter on dry feed they would eat many weeds and plants that they wouldn't otherwise touch.

Photosensitization wasn't confined to cattle, and over the

years I saw many horses have problems as well as lots of sheep. Sorrel horses would get sore legs like the cows if they had white feet and legs, and many times the white blaze on the face would blister and peel off leaving a scar. Gray horses could have problems over the whole body and pintos were like the Holstein cows.

In sheep it affected the white open-faced breeds and could cause some severe problems. Because the sheep eat a lot more browse and weed-type plants they were often involved. On the face where there wasn't any wool the sun irradiated the chlorophyllin and the skin would swell up to three-quarters of an inch thick. It sometimes swelled the eyes shut making them blind for a time. All of the sheepmen seemed to know about the problem although they didn't know exactly what caused it. To them it was an entity they called Big Head. When it would hit the sheep they would try to shade their heads under bushes or one another seeming to realize that the sun made it burn more.

The bright clear air of the mountains gives the sun free rein, and the higher the altitude the less atmosphere it has to penetrate. This made the dryer western ranges particularly vulnerable and the Uintah Basin came up with its share.

Another interesting phenomena if noticed was a darkening of the urine in the early stages. I used this symptom to help in the diagnosis and once in awhile it alerted us to get the animals out of the sun before more serious damage was done.

Forage Poisoning

The other problem, forage poisoning, that seemed endemic to this area was a real enigma and to some extent still is. I don't know that it was my first case of this but it was one I remembered. I went out to a field where the owner's wife pointed out a small steer calf standing all alone. When I got out to him he was apparently blind and so empty his sides almost touched. I felt and looked in his mouth and it was empty, perhaps a little dryer than normal and the tongue just stayed out of his mouth to the side where I'd pulled it. It was a real puzzle and I didn't have any idea what might have caused it, or what to do for it. Something must have indicated to me that he could be thirsty. It was a hot afternoon and I decided to carry him out a bucket of water to see if he could swallow. When I held the bucket up to him he stuck his head down in it clear up to his eyes, but he couldn't drink a drop. I held up his head and poured some water in his mouth. He could swallow just fine, so I slowly poured the water in a little at a time and let him drink it. When the bucket was empty I thought I'd better get him some more, so I started to move. He held his head against my hip and began to follow me. I put my arm over his neck and helped him keep track of me as I moved toward the corral. It seemed like he didn't want to lose track of me after I'd given him that drink. I put him in a little building in the shade and gave him several more gallons of water, then pulled a little green clover and put it in his mouth with my hand. He began to chew and wallowed it around in his mouth and finally got it down. I didn't know of any treatment that I thought would help him, so I told the wife to just pour water down him and put feed in his mouth. In about three days he was eating and drinking again.

This was on one end of the spectrum and the other was a herd of cows trailed for about four or five hours, then turned into a new pasture. The next morning there were four dead and seven or eight in varying degrees of blindness, endless walking, and churning through willow patches, or standing like the calf. There were two down flat but still alive.

By this time I had encountered quite a few individual cases and found the hand feeding and watering about all that could be done. One cow I'd observed for ten days in a prostrate position. When I necropsied her and examined her brain I found some areas of liquefaction. By sending some tissues to the lab in Logan we confirmed the condition as encephalomalacia.

I knew that feedlot cattle had encephalomalacia but I didn't know what was causing it here in this area. I had my suspicions about the plant involved because I'd found it present in about every case. It was a broad-leafed grass, or actually a sedge with a bluish tinge to the leaf, that grew where it was wet. The only problem with my incrimination of the plant was that there were fields of it growing where cattle grazed all the time and they didn't get sick. I had found it present in a problem down the river from Myton where the pastures dried up in the late hot summer and the cattle moved into some swampy old river channels. Similarly one dry summer when irrigation water ran out for the Mountain Home area, cattle were forced into the bottoms and swampy areas there and we had problems. It wasn't always in so-called swampy areas, because I saw cases where the cattle were on good, well-drained pastures, but a canal or other ditch ran through the field and their banks were lined with the plant. It seemed to be the only common denominator in most every case and in most cases there were only one or two animals involved.

Armed with the experiences of the past, I went forth to assess the situation with my good friend Willis Hammerschmidt who had the four dead cows, two down, and half of his herd showing symptoms. We followed a well-beaten trail of fresh tracks leading from the gate where he turned them into the field to a drainage wash with a small stream of water running through it. This is where they had gone to drink after the long drive on a warm day. You could see the fresh stubble of the sedge where they had begun to graze. They had followed up that stream and cleaned the banks for a long way before climbing out to the field on the other side. Special circumstances sometimes play havoc and this seemed to be one of them. If the cattle had drunk their fill, then gone directly to the good feed in the field they might never have eaten the poisonous plant. Fate sometimes turns our luck all bad and this was the case with Willis. With so much loss in his small herd he couldn't make a go of it on his farm and had to give it up and go work in the coal mines. To me it was another strong confirmation of my suspicions about the plant involved and within a year I had two more experiences that pinned it down to that plant. One involved a dairy cow of the Cummings family. They had piped in a spring for water and abandoned a small pond where the cows used to drink. It was fenced off to keep the cows out of it. One morning when he went to milk, Mr. Cummings noticed a cow in heat and being ridden very heavily by the cows in the herd. When he finished milking her he turned her into the dry lot, which contained the pond, to isolate her from the herd and to be handy for insemination in the afternoon. He fed her some hay, but she apparently preferred to eat the green blades of the sedge growing from the now very shallow pond. The next morning I was called and once again I found my toxic plant as almost the only inhabitant in the pond.

The other case became a clincher for me and happened on the Bristol place in Bluebell. There was a ditch that ran through the farmstead and one big corral. It made a turn around a post and in the course of the summer, the earth on one side of the post washed away. The post with a panel of pole fence fell over, covering about fifteen feet of ditch bank. In the fall when it was time to wean the calves they straightened up the post and locked the calves in the corral. The next morning there was a big black calf down, tongue paralyzed, and showing the typical symptoms. When I looked around for the source of trouble, the only spot with any green feed was the triangular piece of ditch bank that had been covered by the pole fence. There was the fresh stubble of my incriminated plant, and I was now sure that it was the one.

Over the years I had been in touch with the Federal Poisonous Plants Laboratory at Logan and had invited them to come to the Basin to help find the source of our troubles many times. I had sent them tissue samples from time to time and tried to learn as much as I could. When I told them I knew for sure which plant caused the poisoning they finally agreed to send someone out. The plan was to collect enough of the plant to feed one of their animals at the lab where they could make an on-going analysis of stomach contents, blood samples, and plant chemistry, with the hope of finding the causative agent.

They had some animals with windows implanted in their stomachs through which samples could be collected and analyzed.

When they came we spent most of one whole day collecting plants from areas I had known to produce the symptoms in animals in the past. We filled large plastic bags with the plants until we had a whole pickup truck full and they took off for the laboratory. Within a week or so they reported that one animal's stomach had been stuffed full of the plant and nothing happened. They fed it to other animals young and old and no symptoms ever showed up. All of our effort and my years of pinning down the plant just seemed to go down the tube. All we could do was speculate that some ingredient in the plant changed rapidly upon harvesting, or it was a mold or something growing on the plant, which might be responsible. It still remains an unsolved mystery.

I have seen the poisoning in cattle fed corn silage to supplement winter grazing, and two horses appeared to have been poisoned by eating this plant in the winter. One horse with the classic symptoms occurred in the late winter when the fields were covered with snow, except for a small area around a warm spring. The warm water coming from the spring kept the ice and snow melted and the plant was growing in the running water.

Some answers have been postulated for those animals affected in the feedlots which show the symptoms, but they don't add up right in the ones eating the green plants. It's still an enigma.

Because encephalomalacia wasn't an easy term for the cattlemen to learn and say, one of my good friends, Mark Bleazard, coined a common name for it from the most classic symptom of a flaccid, partially paralyzed tongue. He called it "Rubber Tongue."

Other forage poisonings that I have encountered have been few in number. They include loco weed poisoning in sheep, one case of greasewood poisoning of cattle, Russian nap weed poisoning of horses, and a few cases of cyanide poisoning from frozen chokecherry plants and frozen corn.

The loco weed poisoning occurred in the early spring, out in the Willowcreek area. The plant was one of the first green things to grow and after a long winter of dry feed, the animals ate it very readily. This is a typical pattern because animals seldom touch it if there are other choices available. The number affected was relatively small, but the plant is well named, because they walked aimlessly about, even walking off from ledges and cliffs like they were mentally out of it.

The greasewood poisoning occurred when a small herd of cattle were taken on a long drive out across a dry barren prairie and into the bottom of Sowers Canyon where greasewood grew abundantly and water was available. Once the cattle had a drink they turned to the abundant green plants to satisfy their hunger. When the owners arrived the next morning several cows were dead and more died during the day. The plant contains large amounts of oxalic acid which crystallizes out in the kidneys, plugging and stopping their function.

The poisoning of horses with Russian nap weed is very rare and only occurs when the animals are forced to eat it because no other food is available. The cases I have seen have been very pathetic in that a paralysis occurs in the mouth and throat preventing them from either eating or drinking. The poor animal about goes crazy with thirst when water is there before them. The first cases I saw resembled encephalomalacia to some extent, and not having access to any research findings, I thought they might respond if I could get water and food into them. I passed a stomach tube into the stomach and pumped in plenty of water and some soluble protein, along with some milk and a gruel made of grain. It did relieve the choking thirst and the animals looked a little better, but the paralysis remained the same. I finally called over to Colorado State University to the vet school and found that once the animal shows the paralytic symptoms, it is permanent, and nothing can be done to correct it. With a bit of disappointment in my heart I euthanized the two horses, relieving them of further suffering.

The weather sometimes causes poisoning that otherwise

wouldn't happen. One spring the cattle were put on Taylor Mountain above Vernal for summer grazing and a severe frost occurred toward the end of June. It froze the leaves on the chokecherry bushes and this made them more palatable to the animals who wouldn't normally eat them. At the same time the freezing caused a breaking down of the sugars in the leaves, which released hydrocyanic acid, a deadly poison to the animals. The combining results caused quite a few cattle to die from cyanide poisoning. It was one of those interesting and challenging phenomenas where you had to look at all the evidence, then put two and two together when other things didn't add up. I added frost to chokecherries and came up with cyanide, a deadly poison.

Another very puzzling poisoning case was one in a pasture up at Lapoint. Glenden Lamb had around seventy head of yearling steers in a field and two of them died rather suddenly. When I arrived on the scene Glenden and his boys told me another one was acting strange. It would be standing normal as ever, then all at once take off on the run and shortly fall to the ground in a convulsion or tonic fit. After a short time it would relax and get back up on its feet, normal as before. I thought I knew what the problem was because I'd seen animals poisoned with chlorinated hydrocarbon insecticides act about the same way. When I ask them about how and when they had treated them for lice and ticks the answer was that they hadn't treated them at all. My next thought was that there must be some can or bucket containing the insecticide in the field. While they made a thorough search of every foot of that eighty acres I walked over most of it looking for some plant that might be incriminated. There was nothing which rang a bell and I was puzzled. I asked them when cattle had been in that field before and if they had had any trouble. They said cattle were in there the summer before as they had been for years and this was the first time any had died. Things weren't adding up for me so I decided to open up one of the dead ones to see if I could find any clues there. As I proceeded through the necropsy everything was as normal as if I'd just killed a healthy animal. I was thinking about removing the head and taking it back to the clinic to remove the brain because of the central nervous involvement I'd noticed in the one still alive. Perhaps I could get it to a lab for some tests. I decided to check out the contents of the stomachs one more time before I took the head off.

I began digging a little bit at a time with my knife and looking at it very closely. A small white piece of something showed up and as I picked it up and pressed it between my fingers it appeared to be a chip of paint. I kept searching and kind of felt like I was grasping at straws but I found another piece a little larger which I showed them and asked if they had seen any paint cans in their search of the field. Not a one. Had there been any trailer or truck left in the field overnight? No they hadn't and didn't think anyone else had. I was about at the end of my rope when someone wondered if it could have been in the other field up on the bench. Then it was explained that the animals had been moved from another field the morning before. They had left a few head up in the other field and they were all healthy, but it did change our object of inquiry. When we got up to the other place there was an old abandoned house and I thought maybe I'd found my source of paint. I sent the others out to scour the fields while I checked out the house. I didn't find all that much evidence of licked off paint and was beginning to question my suspicions of lead poisoning when one of the boys came carrying an old rusted five gallon paint bucket. The bottom was almost completely rusted out and you could see where the cattle had been licking and eating the dried-up paint left in the bucket. Two tiny chips of paint gave us the straw to grasp and we found the answer to a mysterious case of lead poisoning.

Nitrate Poisoning

I received a call from Lewis Farnsworth one spring with explanations that some of his cattle were down as if paralyzed. I hurried up to his place and we drove out into the pasture where two cows were down. I quizzed him about feed and management changes and anything else that might have brought on the problem. The cows had been in the same pasture for two weeks and no changes had been made. He did say that nitrogen fertilizer had been applied the day before, but with it being as widely scattered as it was by the spreader, we didn't think it possible for the cattle to pick up any significant amount. It did give me some food for thought and I decided to attempt some treatment with sodium thiosulphate and methylene blue, the only detoxifying agents we knew of at that time.

When I slapped the needle into the jugular vein the blood came out very light colored like arterial blood. From somewhere back in my memory we had been told at school that nitrate poisoning was supposed to cause the blood to become a very bright red. I was convinced that it was nitrate poisoning and explained to Lewis why I thought so. We treated the cows with the solution in the vein and they showed a positive response and later recovered. When we finished the treatment we began to drive around the field in search of some spilled fertilizer. Sure enough in two places we found little piles of fertilizer where the spreader must have stopped allowing some to leak out. The puzzle was solved and I learned a valuable precaution to pass on to others when they put fertilizer on their fields.

There were a few other cases, mostly where heavily fertilized oat straw was fed, but they were few in number. The most devastating and heaviest losses I ever encountered from poisoning was from nitrogen in the VanTassell brothers' herd. They had wintered one of their herds in a field near Myton, then in the spring drove them from there to a ranch above Duchesne. It was a rather long drive and the cattle had been kept moving steadily all day. When they were turned into the field above Duchesne they were hungry and tired. They fed them a load of hay which they had brought with them from a supply near Myton. They had purchased this hay from Mr. Frank Lidell, one of the better farmers in the area. It was the first cutting from some newly planted alfalfa which was started in a heavily fertilized grain field, and it did contain a considerable amount of the red root weed. The VanTassells had been feeding small amounts of it to the cattle as a supplement to the winter grazing without any trouble. However with the generous feeding and the hungry cattle it combined to cause heavy losses. The next morning after the move there were dead cattle all over the place. I was called and after necropsying several animals I suspected it was nitrate poisoning, although I wasn't very sure. I collected some tissue samples for the laboratory and also some samples of the hay for analysis. There were no more losses after the first onslaught and over the next few weeks it was determined that the red root plant had taken up and contained very high levels of the nitrate. Why had it happened when the hay had been fed previously without any symptoms? The cattle were hungry and the quantity consumed exceeded the critical point. It once again confirmed my conviction that feed changes need to be made gradually.

Anaplasmosis

One day in the middle of the summer I answered the phone.

"Doc, this is Bernard Oberhansley. I got a cow down. Could you come up and treat her for me?" The first thought that ran through my mind was hypocalcemia or milk fever.

"Is it your milk cow?"

"Naw. It's one of my beef cows. I was moving them to another pasture and I noticed she was lagging behind the herd. She acted weak so I left her to follow and went on with the others. She followed a little ways and then just laid down. When I came back by after closing the gates she was still there and didn't offer to get up." My next thought was red water because there was a lot of it in the Basin.

"Do you think it could be red water? If she's bad enough to go down with that I don't think we'd have any chance of helping her. Did you vaccinate in the spring?"

"Yeah, I vaccinated about the end of May when I branded

the calves. I don't think it's that. She didn't really act sick, just weak. She might be dead when I go back over. Anyway, I'd like to know what it is. Could you come up and check her out?"

"Oh, I can come. I was just thinking about what it might be. I'll be right up. By the way where is she?"

"She's in that lane east of the Sundance Grounds. I'll eat my lunch while you're on your way and meet you over there."

My mind was racing around trying to line up the possibilities as I went out to the truck and checked on my supplies. There was calcium; I had seen a beef cow or two go down like the milk cows. I had some sterile water and that new powdered tetracycline, some sulmet, and penicillin. I thought I had about everything I'd need. As I drove along I wondered if the cow might have spinal problems. I'd seen a few cases of paralysis from spinal abscesses and injuries.

It was a warm day with temperatures hovering in the low eighties and it felt good to have my window rolled down and get the breeze as I drove along. I went up past the airport and wound my way up the Todd dugway and through the saddle in the large red buttes north of town. These buttes are the southern exposure of the high flat benches that slope down from the mountains. The multiple layers of red sandstone ledges with the soil in between sets them off in stark relief. It was not unlike many other outcroppings in the Basin that stand as one of its trademarks.

As I proceeded on toward my destination by the Sundance Grounds, I reflected upon this somewhat sacred area to the Indian people. It's a tract of land some two to three hundred acres in size located about five miles from the foot of the Uintah Mountains, sitting on the edge of the flood plain of the Uintah river. It remains in largely its pristine condition except for some attempted flood irrigation from the ditches along its upper perimeter. There is a stand of Scrub Cedar forest in the southwest part of it with sagebrush, rabbit brush, shrubs and native grass intermingled among the trees. In the central area is a large blanket of native grass meadow studded here and there with clumps of tall brush and brambles of many varieties, the bull berry and buffalo berry being in abundance. It is in this grass meadow where the sacred dances are performed and where the Native Americans camp while attending the dance ceremony. Their teepees, tents and wickee-ups spring up by the hundreds, scattered among the clumps of foliage. (The wickee-ups are small bowries made from poles with fresh-cut, green, leafy tree limbs woven over the top and down three sides making a cool escape from the burning sun.) It's a thrilling sight to drive by and see this great celebration.

When I drove down the lane it was lined with willows and tall brush with black willow trees here and there. The old cow had picked a grassy spot on the west side in the shade to lay down. She didn't try to get up when my truck approached, but to look at her it appeared that she was normal. Bernard was waiting on his horse. I got out and walked slowly up to the back end of her, spit on my thermometer and inserted it into her rectum. She didn't offer to get up. I parted the lips of the vulva and the membranes were very pale and white. It was my first clue that something indeed was wrong. I gently stroked under the vulva flicking the end of it with my fingers trying to stimulate the old gal to urinate, but it didn't work. A sample of dark bloody urine would have helped me to know if it was red water, but the very pale membranes were not like the muddy icteric ones I'd seen with that disease. Over the years I did learn that in the acute stages of anaplasmosis the urine frequently became dark and copper tinted. I pulled out the thermometer and found a mildly elevated temperature. The thought came that she might have heart trouble, so I moved up in front of her to check on the jugular veins to see if they were distended and had a pulse. When I leaned toward her she lunged to her feet and took after me with the apparent intent to do bodily harm. With a little fancy footwork and a dash around my truck I was able to escape. With that kind of incentive I frequently found that I could move quite rapidly. She walked a few yards up the lane and began to shake and wobble on her feet. Then she turned and faced us shaking her head up and down in a threatening way.

The lunging to her feet, raising her tail, and charging me ruled out the spinal problems, and she didn't show any heart symptoms, and she wasn't typical of terminal red water. Maybe it was the adrenaline she put into my brain that made me think of anaplasmosis. When I'd studied about anaplasmosis in vet school, the picture of the little dark inclusion bodies in the red blood cells made a deeper impression, I guess, than the word picture of a severely anemic cow. I don't recall that we ever got to see an acute case, but I knew it was out there and this must be what it looked like.

"Bernard, I think this is a case of anaplasmosis." He had stepped off his horse and tied him to the fence and was leaning against the front fender of my truck.

"Am-plas. What'd you call it. I ain't never heard of that before."

"A-n-a-p-l-a-s-mosis. It's a malaria-like disease that breaks down the red blood cells. Kind of a blood parasite. We'll have to treat her in the vein for it."

"Can you save 'em?" It was a natural question but somehow it troubled me a little. When you're dealing with sickness and life, who could answer that question. In this case I hadn't ever encountered it before. I didn't have any experience to call upon.

"Well to tell you the truth, Bernard, I don't know. This is my first case, if that's what it is. We studied about it and they say it responds to this new drug called Terramycin. I've used it on pneumonia and scours some and it seems to work. I think we should give it a try."

"OK, I'll get my rope."

The old gal remained quite belligerent mentally so Bernard got on his horse and undid his rope. When the rope tightened on her neck she pulled back for a second then flopped over on her side and I thought she was going to die right there, but she didn't. In a minute or two she rolled up on her brisket and just laid there sort of resigned to her fate. I mixed the powder in the sterile water and put on the IV set. When I approached her again she just stayed put and I popped the needle in the jugular and gave her the medicine. When I saw Bernard at the sale the next Saturday he said she was up and eating, seemed to be alright.

I didn't take any microscope slides with me that day to make a smear and look for the diagnostic inclusion bodies, and I wished I had, but the severe anemia, the watery blood, overall weakness and a slight yellowing of the white skinned areas always stayed in my mind. Over the years I saw many cattle with anaplasmosis and for a while tried to make smears to see the inclusion bodies. My success with the microscope was so seldom I eventually gave it up and relied on the picture in my mind of Bernard's cow. The belligerent attitude seemed to be caused by anoxia of the brain from the severe anemia. I saw it in other anemic conditions as well.

I think the disease came into this area about the same time that I came back to practice here, because it increased steadily until the more effective parasite control medicines became widely used. It is a disease transmitted by insects, especially the blood-sucking varieties like mosquitoes, flies, ticks and lice. Some of the animals that have it recover and become hidden carriers, because they show no symptoms. Hypodermic injection needles have been shown to carry it from one animal to another as well.

The incidence is higher in the summer and in pastured cattle as opposed to those that summer on the mountains. I have seen it in the winter and presume it is transmitted by the ticks and lice then. It can be a more serious problem when it is cold because I have seen tails and ears frozen off. In a few severe instances the hooves have frozen and sloughed off as well. I don't know whether some areas have more flies and mosquitoes than others, but I have noticed certain areas have more anaplasmosis, and some herds more than others.

Arden Evans summered his cattle east of Upalco and he had many cases over the years. With the development of the tetracyclines he was able to control the losses and the incidence diminished as well. I think his diligent and rigorous treatment eliminated a lot of the carriers.

One year in the late spring Frank Arrowchis had four cases of anaplasmosis in the space of a couple of weeks. He ran his cattle on the river bottoms of the Whiterocks and Uintah rivers and it wasn't always easy to keep a close watch on them. When these four cases showed up he became alarmed and wondered if there was a way to eliminate it from his herd. Fortunately the state diagnostic laboratory had just geared up to make a blood test on cattle. I explained that we could identify all the cows that had the disease and they would contain the carriers. He thought he was willing to make the sacrifice and sell those that tested positive. We arranged for Richard Olsen, the federal technician in the area, to do the bleeding and send the blood in for a brucellosis test with a request that it be tested for anaplasmosis as well. When the results came back about ninety percent of his herd tested positive. He was devastated until I pointed out that almost all of his herd were immune and his troubles should diminish rapidly. There was a vaccine available, but it was so expensive most owners declined to use it.

The Grunts

The very costly condition of pulmonary emphysema and edema got its common name of the grunts from the symptoms exhibited by the cow. The dominant beef cattle enterprise here in this area was the cow-calf operation which utilized the mountain ranges for the summers and then wintered the cattle on the ranches. Because the winters were cold and the feed limited, most operators raised the calves for sale as weaners in the fall. This management practice required some drastic feed changes when the cattle were shifted from the ranch to the range or from the range back to the ranch. The problem usually arose when the cattle were trailed from the dry ranges to the lush feed of the ranches. They arrived tired and hungry and ready for a good meal. When they ate the lush feed with a higher quality protein, one of the amino acids, tryptophan, was more abundant. As tryptophan metabolized, a toxic material was released which paralyzed the tiny muscles which opened the air passages and caused fluid to accumulate. If the emphysema and edema were severe, the cow would be unable to exchange sufficient oxygen and carbon dioxide to survive. When the animal began to come down with it, or had only a slight case, the cow would stand with head extended and with an abdominal, compressive type of breathing, with a grunt accompanying it. Thus, the name the grunts.

There were some medicines that helped relieve the animal. However, the exertion of moving or handling pushed them beyond the point of survival, so it was most often counter-productive to attempt treatment. In animals that were not too wild I used a "have gun will shoot" method where I pulled a shot of medication into my syringe, then moved slowly around through the herd. When I saw a sick one with its head turned the other way I would dash up behind her, jab and squeeze in the shot before she could move. It was difficult to evaluate how effective my treatment was because there were all different stages and degrees of the disease, but most of the owners felt that I had helped them.

If treatment meant driving the animal more than a short distance on a slow walk, or having to rope them I always advised leaving them alone. My jokingly given advice was to take a trip for ten days and come back and count the dead ones. The grunts was a very serious condition causing death losses, and additional animals came down with it for ten to fourteen days after the first onset in a herd. The severity I observed was directly proportional to how hungry the animal was and how lush the feed was that they were introduced to. One could predict it happening in a lot of cases, but there were some outbreaks that occurred that I didn't expect. I always felt that it only took one bellyful to bring it on, while at the same time cattle became affected up to ten or fourteen days after the first onset, indicating a continuous consumption problem.

Typically the cattle were put into the field and the first symptoms started showing on the third day if it was severe. Then additional animals came down with it up until the tenth day, the majority occurring on days three, four and five.

Correspondingly, the death losses were higher in those that occurred in the three, four, and five day range of onset. The average number of cattle affected in a herd was about thirty percent with death loses ranging from three to ten percent of those affected. There was really no way to tell how many might get sick or how many might die. Sometimes I would be called out to necropsy one dead cow in a herd finding that it was the cause of death, yet only a few others showed very mild symptoms on close observation.

Over the years there were hundreds and hundreds of cattle lost to this condition. I felt it was the greatest cause of loss in the beef cattle of the area. The variations of outbreaks and the conditions setting them off were limitless. Let me outline the extremes by two separate occurrences.

The first and most unexpected occurrence was in twenty or so bulls at the Lusty ranch. They were kept in a small pasture and were being supplemented by feeding hay each day. The second crop hay was hauled in from the field. It was the end of the day when everyone was tired, so Clark decided to throw the bulls their few bales of hay from the new crop instead of unloading it and then reloading some from the old stack. Three or four days later I was called to come and check on some sick bulls. When I walked out through that bunch of bulls I could see the extended muzzles and just a slight labored breathing in about a third of them. One might have thought they were getting pneumonia, but my experience told me that a third of them wouldn't get it all at once. I concluded that it had to be emphysema and questioned Clark about any pasture change or feed change. At first he said there hadn't been any, then he remembered feeding them the new hay. This was the only change and only one feeding, yet it brought on the condition. Fortunately, they were very mild cases and soon recovered on their own, but it proved my theory that one bellyful could cause it.

The most severe outbreak I ever encountered was in a herd of cows that spent the summer on Mosby Mountain. That fall they were driven from there down across the Duchesne River by Randlett and up to a ranch on the south side. It was a long twenty mile drive with very limited feed on the way and the cows arrived very hungry. The owner was new to the area and not aware of any danger except bloat. He had alfalfa fields with ten or more inches of growth on them, but it had been frosted so he took a chance on the bloat and turned them in. He watched them very closely for bloat and didn't have any trouble with it. He had about one hundred and twenty-five head and when I was called about four or five days later there were already ten head dead. I opened one of them up to show him what the trouble was and explained how it happened. As I looked over the remainder of the herd, I estimated there were about seventy more of them showing symptoms, some severe and some just starting. I was just sick knowing that he was going to loose a lot more and he didn't even know it. It was the fifth day and the feed was well topped so I told him to leave them in the field where they were and leave them alone, there wasn't anything he could do that would help them. I told him if he tried to move or treat them more of them would die than if he left them alone. He was just a young man getting started and I knew this was going to wipe him out. When the losses finally ended, the by-products man told me he had picked up thirty or more head, and many of those that survived were walking skeletons by spring. He lost the ranch and all of his equipment and left the area.

Another experience convinced me that it could happen in sheep as well. The Mortensen brothers planted and fenced their farm so they could rotate pastures with their farm flock of sheep. It appeared to be an ideal arrangement where the sheep could be pushed on new pasture every few weeks and control the internal parasites at the same time. The only trouble was a high rate of respiratory problems and some losses. I was convinced it was an emphysema-edema related problem and when they stopped rotating, the respiratory troubles cleared up.

Everyone knew or soon learned that prevention was the only answer and many methods were utilized to try to avoid the sudden feed changes. Putting the mountain cattle into partially grazed pastures for a few days before entering the more lush feed helped some. Another method was to feed them hay for a few days prior to going into the fields, and especially to fill them up on hay just before turning them in. As we experimented with various approaches, I became convinced that a gradual breaking in was the only safe way. If facilities were available, and many times they weren't, I would recommend that the cattle be put in the field for one half hour the first day, then taken out. After about six days of increasing the time by one half hour each day the cattle could be left in the pasture. If the cattlemen were diligent about putting them in and out as recommended I never knew of an outbreak.

As I mentioned earlier, some of the cows that had emphysema and survived were prone to other problems later. Some died with pneumonia, others would develop brisket disease the following summer on the mountain, and some just stayed poor and didn't do well. Those that went back to the higher altitudes with damaged lungs placed an extra burden on their hearts, which caused them to fail. When I would necropsy some of the brisket disease cases I would find large open spaces in the lungs from emphysema earlier.

Pulmonary emphysema/edema didn't just happen in those cattle that went to the mountain ranges. The pasture raised cattle were plagued by it also. Here it occurred primarily in the spring when they were put out in pastures that were more lush than the ones they'd been in. It could happen at any time of the year when sudden changes were made. Another problem was a breechy old gal that would crawl through the fence and get into the alfalfa field. If the bloat didn't get her, the emphysema sometimes caught up with her. No matter how it happened it was a serious problem. When I'd get a call and the owner would say, "Doc, I think my cows have the grunts, could you come up?" I'd feel a little sick to my stomach and feel sorry for my poor struggling friend.

Red Water

Not long after I came to the Basin to practice I was faced with the problem of red water disease which was taking its toll and spreading. I knew that immunization was the only way to control and limit the losses and I kept a good supply of vaccine on hand. We had the sulfa drugs and penicillin but they were not much help in trying to treat this disease. I knew from school that liver fluke played a role in the disease, so I could advise and explain its presence and importance. There were many who tried to treat for fluke thinking they could stop the red water, but I knew that it was the migrating larvae that caused the liver damage where the red water organism grew, so killing the adult fluke didn't always protect them.

One day I got a call from a fellow who knew that his neighbor boy had shot one of his cows. He just needed the proof that the cow had been shot because he'd seen the kid hunting over along the river where the cows were. He asked me to come up and open up the cow, then he'd call the sheriff and make them pay for the cow. He knew he'd seen those cows a day or so before, and that kid was shooting down in there, and he knew what had happened. He said the kid had shot her in the left eye because it was put out and all bloody, also a little blood was coming from the cow's nose. When I got there the magpies flew up into the trees and I could see they had been eating in that eye socket. We rolled the cow up on her back and I began to skin her out and open her up. He railed on about that kid that's always hunting. He was talking all the time until I showed him the necrotic infarct in the liver, and the blood-red urine in the bladder, which I explained was an unfailing sign of red water. Things got very quiet after that and he just asked me how much he owed me.

There was one drug company, Pitman Moore, that had an antiserum. I obtained some and tried it as a treatment and I found it was quite effective if used early. When I received a call and the animal still had a high fever, one bottle of the antiserum given in the vein, along with some antibiotics, would usually cure them. It was a bit expensive, but it saved a lot of cows and I was pleased with it. The serum was made from a hyper-immunized horse and had a cautionary label warning about antiphylaxis, but I never had any reactions with it. All at once the supply of antiserum ceased to exist. After several long distance phone calls I found that the one horse which produced the whole nation's supply had died of colic. It was determined by the drug company that there wasn't enough serum sold to justify production so another horse was not hyper-immunized.

I was very disappointed and tried to get other companies to make the serum. I even offered Utah State a horse if they could get the process from Pitman Moore and make up the serum. Since that time most cases of red water have died. The carticosteroids and massive doses of antibiotics have saved a few animals.

From years of observation I came to the conclusion that weather conditions played a role in the spore-forming clostridial diseases. Whenever the condition of cold nights and hot moist days existed I could expect to see more cases of red water and Black's disease. This was true in the spring and fall, but more prevalent in the fall when the nights first start getting colder. I also learned that the vaccine would only protect the animals for about thirty to forty-five days if given just one shot without any previous sensitization. On several occasions cattle from other areas of the state had been brought out here to pasture and given one shot of vaccine at the time of transport. In thirty to forty-five days after arrival, they had died with the disease.

Black's Disease

I had dealt with red water for a number of years and had diagnosed a few cases of Black's disease in some farm flocks of sheep. I got a call from a young fellow who was just getting started in the dairy business. He had about thirty head of two and three-year-old heifers. He had remodeled a barn and worked at another job, like a lot of others, to support his family. One morning when he went out to milk there was a heifer dead. He called me and explained that he had milked her the night before and hadn't noticed anything wrong. My mind raced through the possibilities and the only thing that fit in my thinking was bloat. I quizzed him about his feeding and how much and what kind of grain, (rolled barley), he fed. I suggested maybe she had just bloated, but he didn't think she was that tight. He had to go to work but he asked if I could come over and open her up and see what killed her.

There wasn't much that I could find that told me what had been the cause of death, but I was satisfied it wasn't bloat. The most significant thing was some inflammation in the connective tissue between the skin and the muscle on one front leg and up along the neck. I concluded that it had to be blackleg. When I talked to him at work, he said they had been vaccinated as calves, but not re-vaccinated since then. I recommended that he vaccinate for blackleg and he stopped by and got the vaccine on the way home from work. He vaccinated them all as they came through the barn and we thought the problem was solved. About a week later he found another one dead that he'd milked twelve hours earlier. When he called me I was puzzled because animals were usually sick a little longer before they died with blackleg. I assumed that the vaccine hadn't had time to work. Then one afternoon about ten days later he called me and said one of his cows was sick. Could I come over?

When I arrived the heifer was definitely sick. She was out in a big corral. We started to drive her around and over to the gate leading to the barn. When she got under the shed she laid down and while we were watching, rolled over on her side and died without a struggle. Something was killing off his herd and I didn't know what it was. It wasn't the blackleg I had suspected, because it was two weeks since they had been vaccinated. I was really puzzled and felt a burden of responsibility to find the answer, for him and all the other cattle owners in the Basin. I decided to do a very careful necropsy and get some tissue samples into the diagnostic laboratory. What I found was much the same as with the first cow with perhaps a little more pronounced inflammation. This time the heart sack was filled with a sanguineous fluid. I took a lot of samples and rushed back to my office to refrigerate and prepare them.

The submission of samples to the diagnostic laboratory in Logan presented a real challenge because it wasn't possible to keep them refrigerated for the two and three days it took for the mail to deliver them. The putrefactive organisms would overgrow the pathogens and even alter the tissue structure. We tried various agents like borax to keep things dehydrated, but our success was very low. I called the lab for their recommendations and learned that someone in the extension service was out in the Basin. I checked locally and found that they were going back to Logan within the hour, so my luck was running high for a change.

Sometimes when you don't know the specific answers, your general knowledge tells you to look at the basics, look for possibilities. This prompted me to take a few minutes and look around his setup while I was there. He had a water trough that was running over and across the lane where the cows came out of the barn. While I was looking around one of the heifers came up the lane and began drinking out of the puddles made by the cows' feet. The water wasn't clean or fresh and was seeping through the manure making a big wet area twenty or thirty feet in diameter. When I saw the animal drinking from the small puddles instead of the water trough, a red flag went up in my brain. When I couldn't find anything else to suspect I advised him to close that area off and route the cows back across the concrete to the big corral and keep them out of the wet area. That turned out to be a very important recommendation.

I didn't hear from the lab for a week and a half so I called them. They said they had obtained some good cultures that appeared to be clostridium organisms but to identify which ones would require running them though the various sugar medias, taking up to three weeks. I harbored a bit of worry and uneasily waited for a call to tell me another cow had died but none came.

About a month after I had sent the samples I went up to

Logan for a continuing-education seminar. The first day after the meetings, I dropped into the lab. They said they had just finished reading the cultures that day. The girl went to get the report. When I looked at the results and saw clostridium novyi, I didn't hardly believe it. That was Black's disease and my information indicated that it was only a sheep disease. I wondered if the samples got mixed or something. I checked with the doctors and they assured me that it was indeed clostridium novyi and it had been reported in cattle. They had isolated it from a cow in Wyoming, about a week earlier than my case. Black's disease in cattle had been confirmed in the Uintah Basin. Now we had to add another vaccine to the recommendations.

Sinus of the Withers

This condition probably had more names, generated more remedies and treatments than anything but lameness in the horse. It was most commonly known as thistalow or fistula and every horse trader and local expert had their own favorite concoction to put down into these draining abscesses to burn the infection out. It was an enigma for hundreds of years and the real cause was not discovered until about a decade after I got out of vet school.

Anatomically the horse and other animals have a thick band of yellow, elastic tissue that runs from the withers or top of the shoulders to the back of the head. This elastic rope-like structure helps to hold up the weight of the head, relieving the neck muscles, and helps in lifting the head when eating or drinking. On the back end where it leaves the withers there are long vertical spines on the vertebrae that it rides across. When the animal put its head down this strong elastic band pulled down across these spines. In the wisdom of the Creator, He placed a synovial bursa or sack-like structure filled with a slippery fluid over each of the spines. They provide lubrication and mobility for the very limited movement of the cord.

It was thought that the horse in going under a tree limb, a

shed or anything similar, would bruise these bursae causing an abscess to form. It began small but rapidly enlarged to a huge swelling, up to half the size of a basketball. They usually swelled out to one side or the other and were very tender and painful. If they were not lanced they would eventually break open and drain, leaving a deep tract down to the bursa and the drainage would never clear up. This deep tract was the recipient of the many harsh treatments which were poured in or injected to "burn out" the infection. Most of the treatments never did cure the condition, only created a lot of scar tissue. Once in awhile the skin would heal over for a time and the medicine man would get credit for a cure. The only problem was that the organism would begin to grow again and it would break open later.

When I was in school, we were taught to do a rather drastic operation where we made a large triangular opening and cut out all the necrotic or dead tissue including the bursa itself. To close it up we put a warm pack of boric acid held in place with gauze sutured at the three corners. Then in a few days when the danger of hemorrhage had passed we removed the pack and let it heal as an open wound. It was rather drastic surgery and much blood was lost but most animals healed and didn't have any subsequent problems. I was only called upon to do a few of these operations.

It was finally determined that the infection was the brucellosis organism and that it could spread to other animals. When we first learned this, very heavy doses of the Strain 19 modified vaccine, which was used in vaccinating calves, was used as a treatment. The results were quite good as far as the horse was concerned, but because of the possible spread to cattle it was soon discontinued and we were instructed to send all infected horses to slaughter if we could convince the owner of the danger. Brucellosis was being eradicated in cattle because of the danger to humans who got undulant fever from this organism. Over the years, as we completed the eradication of brucellosis in cattle and other animals, I never did get any calls to come out and treat a "thistalow."

9
CYCLIC MALADIES

Stomatitis

It was a hot summer afternoon when Nancy answered the phone then handed it to me. "Hello, this is Dr. Dennis."

"Dr. Dennis, this is Lujean Winn. We've got some real sick cows. Could you come down and take a look at them?" Real sick, I wonder what that means.

"Have you got any dead ones?"

"No they haven't died yet, but they won't eat or drink a thing so they're gonna die if we don't do something. There's only two of 'em sick, but all the rest might get it." From her anxiety I could tell it was something out of the ordinary, something they hadn't dealt with before. I wondered what it could be; maybe that forage poisoning where the tongue gets paralyzed.

"Are they wandering around acting partially blind or have a twitch to their head?" I quizzed, hoping for some clue.

"No, they just stand out there in the pasture and won't eat." I knew their pastures had some wet areas where that poisonous plant could grow. I thought it had to be encephalomalacia.

"Okay, I'll come right down. I'm not that busy today." The puzzling circumstances of the paralytic forage poisoning rotated in my head as I started for their place south of Fort Duchesne. It was a warm day and I welcomed the feel of the air as it rushed up my sleeve and across my wet back. If I held on to the top of my outside mirror the fifty-five mile an hour wind had a straight shot right up my short sleeve. By leaning forward slightly the evaporation cooled my whole back. Driving along was a cool respite from the heat of the day.

When I pulled into their yard, Ralph was there and he directed me down over the hill and out into the large pasture. The feed was getting picked over, but there was still some grass available. It was a typical time when cattle began to search for other feed, and that's when they usually got poisoned. I wasn't thinking of anything else. As we approached the first cow she turned her head to look at us and seemed to see alright. She was so empty her sides looked like they were touching each other. Her nose was dry and peeling and the nostrils had a lot of mucus dried around the edges. Cows normally lick their noses and keep them clean, so my first thought was that it might have something to do with the tongue. She wouldn't let us walk up to her, so I got my lariat and threw a loop around her neck. She was surprisingly strong for as thin as she looked. I snubbed her to the ring welded on my bumper and worked her up close to the truck. "Here, Ralph, hold this while I get my nose tongs in her nose. We're going to have to look at her tongue and down her throat."

"I guess that's a good place to start. She sure ain't eaten' anything." I grabbed her nose with my tongs and threaded the rope through the small loops at the top of my truck bed. Once I had her head pulled up I took hold of her upper lip on the far side and slipped my other hand into her mouth from the side, staying cautiously forward of the cheek teeth. I'd learned the hard way that a fellow doesn't want to let a finger stray in between those grinders. As I grasped the tongue to pull it sideways out of her mouth all of the mucous membrane on top of the front half slipped off in my hand. It was painful and she opened her mouth and I could see the red raw tongue and other places in her mouth where the mucosa was blistered and white. I'd seen small areas of necrosis in calves with diphtheria infection but nothing like this. I was a bit shocked.

"I can see why she doesn't want to eat. Did you see that tongue? It was plumb raw."

"I could see it a little. What do you think caused it?" That was a question already traveling at high speed through my brain, I'd never seen anything like it before. My thoughts flipped back to vet school and discussions on foot and mouth disease, and then there was stomatitis and something else—African something or other. It was too many years ago to remember. Maybe she'd just got into something acidic or caustic and it had burned her mouth.

"Ralph, is there any chance these cows could have been licking on some old batteries or been into something caustic of some kind?" I said again trying to find some cause or another.

"Not that I can think of. They can't get up around the machinery and I don't think there's anything in this pasture."

If it didn't have a simple explanation like that, maybe it was one of those exotic diseases. I'd sure have to do some reading and studying when I got back to the office, but I was there. What could I do for the poor cow?

"Ralph, I haven't ever seen anything to equal this. I've seen some sores in cows' mouths before, but nothing like this. I'm gonna have to bone up, hit the books." I didn't want to mention anything about foot and mouth disease. It seemed like a casual remark about something like that could spread like wildfire. By the time I got back to the office, the country could have a full blown epidemic. "I think it might be something we call stomatitis, that means inflammation of the mouth, and she's sure got that."

"What do you do, just let 'em die?" There was a touch of anger and exasperation in his voice. I guess I couldn't blame him. I wasn't much help but my foggy memory told me that there wasn't much if any death loss from it.

"It's a little foggy in my mind, but if I remember correctly there isn't any death loss from it, or at least not many." I thought that might take the edge off of his feelings, but he still wanted help. What could I do? I turned my thoughts back to the poor old cow. Was there anything I could do to help her? I remembered that when I skinned my knuckles or got a cut I always put iodine on it and it seemed to reduce the soreness. Iodine does have the property of killing nerve endings. "I don't know whether it will do any good, Ralph, but I think we ought to swab her mouth with iodine. It might kill a few of the germs and maybe take some of the soreness out to help her to eat and drink a little.

"Yeah, I think that might help her a lot." Doing something, whether it helped or not was always the owner's wish. I put some gauze in my long handled sponge forceps and soaked it with tincture of iodine and began to swab the raw places in her mouth. When I rubbed those other white blistered areas the mucosa came right off. It was pathetic and the old gal didn't appreciate it, but it was over fast, limiting the pain. I turned her loose and focused my attention on the other cow.

The other cow was a carbon copy of the first one and within the week there were more cases in other herds which told me that we had an outbreak of something. It fit the description of vesicular stomatitis which was thought to be a viral disease, although all of the answers were not yet in on it. I read everything I could find about it from my journals, from the textbooks, and my notes from school. Because it was a reportable disease, I notified the state and federal veterinarians and learned what I could from them. In the end the best lessons came from seeing the sick animals and following them through to recovery.

We had upwards of a hundred cases in the course of that season and I don't remember any dying. The cows would starve themselves completely for a week or more before trying to eat, then gradually recover. I didn't treat all of those I saw with iodine, but I felt it did help some of them to get back on their feed sooner. There weren't any death losses from it, but it was a substantial economic drain to the owners because of the severe weight loss and the resulting milk loss to the calves. After that summer and a few cases the following year it disappeared for about fifteen years or so. The next outbreak was diagnosed in New Mexico and Western Colorado and spread from there to our area and on into Idaho. The lesions created were pathetic and mostly confined to the mouth, although some teats and udders were involved especially up in Idaho. There were a few horses that became infected and they ended up with sore and blistered feet. With this disease it was a matter of making a diagnosis and then feeling sorry for the poor animals and their owners.

Brain Fever (Encephalomyelitis)

My first encounter with brain fever in the horse occurred many years before I even thought about becoming a veterinarian. There was a severe outbreak here in the Basin while I was growing up. There wasn't any mention of different strains then, it was just brain fever and it was very severe. There were hundreds of horses that contracted it and nearly all of them died. The symptoms of the horses I saw with it were profound. They would stagger into a fence or corner of a shed or barn, then stand and push with their head until they finally went down. After they were down on the ground they would thrash with their feet and legs and pound their heads violently on the ground. They didn't live too long after they were down, maybe a day or two, but it was long enough to cut and beat their eyes, lips, and face into a horrible bloody mess. The only recommended treatment was to put cold packs on their heads to reduce the fever in the brain. To accomplish this was difficult at best with the horse pushing and thrashing around, but owners would tie burlap sacks on the horses' heads and then pour cold water over them. Some even tried crushed ice in sacks. The water always made a puddle of mud where the horse's head was and they would get plastered with it. The picture in my mind was a horse with a mutilated head covered in mud. Many of them were shot by their owners when the mutilation started.

There were a few horses that did live through the disease, but they were never the same afterward. There was so much brain damage that they appeared to be in a stupor, or they walked unsteadily on their legs. Some appeared to have forgotten all they had ever learned. Some owners kept them around for sentimental reasons, but most were killed or sold for fish food. On our farm we were extremely lucky and none of our horses ever got it.

When I began to practice here, outbreaks seemed to come in cycles. We'd have cases one year, then not see any for several years afterward. Since there wasn't any successful treatment, the beginning of an outbreak caused a panic to get the animals immunized. On the trouble free years nobody bothered to vaccinate and I could not afford to keep any volume of supplies on hand because they became out-dated and had to be thrown away. When the first case was reported, usually in late summer, everyone wanted vaccine immediately. I would call the pharmaceutical supply houses and their supply would be limited. Sometimes they would have vaccine in other areas of the country or the world and have to ship it in. The cyclic pattern resulted in a vicious circle situation and we were all in a panic.

At first the vaccine was an intradermal type and few owners even knew what that meant. Fewer still knew how to give the shot properly. Intradermal meant that the vaccine had to be deposited between the two layers of skin. Placing it there gave it a longer antigenic response time because there isn't an abundant blood supply to that area. This was required to get a good immunity, and it also had to be repeated in two weeks. It wasn't an easy thing to do. You had to slowly push the needle into the skin on an angle and inject a small amount to see if you were in the right layer. If a small weal or lump started to bulge up the outer layer you knew you were in the right place. Now horses are a lot like children when it comes to shots. With a strange person clipping off hair and applying alcohol and then slowly pushing a needle into their hides they panic easily and often did. Sometimes the first shot went fine, but when it came time for that second one it was a different story. I wasn't called upon very often to give the shots because they weren't about to pay a veterinarian to do something like that. I'm sure most of the owners just injected the vaccine under the skin. It wouldn't develop much immunity because it was absorbed too quickly, but it probably gave a little boost. By the time the second dose was given, we'd have a killing frost eliminating the danger from the mosquito.

I do remember one season when the outbreak started at the end of July and before it was over I had administered seven hundred and twenty doses of the vaccine. It was certainly a great improvement when they developed a subcutaneous vaccine that could be given underneath the skin. The owners could truly do their own then. When the newer vaccines came out with tetanus, flu and other diseases mixed in, it was even better because these were given every year and we had a larger immune population. When an outbreak did occur I would go out and collect blood samples to send in and identify which type of encephalitis we were dealing with. This monitoring of cases also helped the pharmaceutical firms to know how much vaccine of a particular strain was going to be needed. After the anti-inflammatory drugs came out I also treated some cases with it, but it was hard to evaluate its effectiveness because of the great variation in severity of cases.

One of the last cases I had was out to Rip West's place by Montez Creek Reservoir. He had a beautiful two-year-old palomino stallion starting to show the symptoms. I took the blood sample and gave him some medicine the first day, then stopped back there two days later. His symptoms were a little more severe, but he was still on his feet. There were three of Rip's neighbors there watching the horse as he would start to doze off then suddenly awaken and adjust his balance on his legs. I took his temperature and checked his membranes. "I think he's got a pretty good chance of making it, Rip. He still has good balance and fair alertness when he comes to after dozing."

"Will he be any good if he does make it?" This was another of those questions that nobody could give a definite answer to.

"Well, I can't give you an answer on that. Only time will tell, but the cases we've had in the last few years haven't been too severe and most horses have come through it fine."

"Where'd he get it?" one of his friends asked.

"It has to come from a mosquito bite and not all varieties of mosquitoes can carry it. Only the Culex variety."

"I thought mosquitoes were all the same. I didn't know they had varieties. They all bite the same and look the same to me. Maybe some are bigger than others," came from another of the neighbors.

"Can my other horses get it from him?" Rip asked and then asked another question as he thought about the mosquito angle. "I mean could a mosquito bite him and then go bite one of my other horses and cause him to get it?"

"No, first of all they have proven that horses can't get the kind we have here in the U.S. by contact with each other. And they have never been able to transmit it from one horse to another by mosquitoes. It has to come from an intermediate host."

"What's that?" the first neighbor asked as he eyed me critically, like I might be telling a big story.

"It's another species of animal or bird. Actually we probably don't know all of the carriers. Birds and the common water or garter snake have been incriminated but may not be the only ones. They do know that it has to come from the intermediate host through the mosquito to the horse or to man. Man does get the disease like the horse."

"You mean we could get it from a mosquito bite?"

"That's right and I guess if horses are getting it we ought to avoid the mosquitoes as much as possible."

"How's a guy going to avoid mosquitoes? Man, when I go

out to change my water ya can't hardly breath for 'em."

"Yeah, I know the feeling. You do have to change water and work out in them, but I guess the better part of wisdom would tell us to do what we can. Wearing long sleeves or a coat and using repellents could help."

"Have there been any people in this area get it?"

"Well, there has been a case or two in the Basin, but none recently that I'm aware of. I wouldn't get all shook up about getting it because the incidence in man is very low compared to the horses. Horses probably get bit by a thousand mosquitoes for every one that bites man, but it is possible to get it."

"What does it do, just affect the brain?" Rip poked his chin in the direction of the horse.

"That's essentially true. It's a virus that gets into the bloodstream and is then carried to the brain where it localizes and grows in the nervous tissue. It produces a severe inflammation with swelling and some nerve damage along with a high fever. None of the antibiotics are effective against it so the animal has to develop antibodies to fight it off."

"I guess that's why they call it brain fever."

"Well sometimes it's called sleeping sickness, especially in man, but if you notice, that horse is depressed or in a partial stupor. As he stands there he starts to go to sleep, his head starts to go down like maybe he is going to fall forward, then his balance organs take over and wake him up. He takes a step or two, regaining balance and then it starts over. You can also tell there is some nerve paralysis because his lips are hanging down without any tone in them, especially his bottom lip. It's in the brain all right and the rest of him seems okay, except he has a high fever. I don't think you will ever see a more typical case than him." I don't know what he may have been dreaming of, but I hadn't any more than got the words out of my mouth when his penis slid down out of the sheath and came up in a hard erection. That's when Wes Grandsen, one of the older neighbors there, eased over beside me and said behind his hand, "Where could a guy get a hold of one of those mosquitoes, Doc?"

10
GUARDING THE MILK SUPPLY

The Metamorphosis of Dairy Farming

Farming and ranching have always been a family oriented effort and this was doubly true in the Basin where it was a struggle to make ends meet. I found the dairy farming enterprises even more family centered than other endeavors. The children, both girls and boys, from a very early age, started helping and working alongside of Dad and Mom to feed the calves, move the cows, and help with the milking. It was an ideal setting for the whole family to learn how to work and to feel the satisfaction of work itself. It seemed like working together for a common purpose produced outstanding young people and close family ties. It was a pleasure to see them grow and mature over the years.

When I first came to the Basin to practice it was reported that there were about one hundred and five dairy farms in Duchesne County, and nearly as many in Uintah County. It was a way of life for the small farmers: a way to bring in a little cash to live on. It started out as a cream check. The cows were milked mostly by hand. Then the cream was separated from the milk with the old fashioned centrifugal bowl cream separator which was turned by hand. When the revolutions per minute reached a high enough level the milk was turned on and out came a stream of skim milk and a smaller one of cream. The skim milk was a by-product and was used to feed young calves and almost always a pen of pigs. For those with too few cows to afford a separator the milk was strained into large flat pans and placed in a cool place to let the cream rise to the top. That cream was skimmed off with a large spoon and the skill of an artisan, then kept in a closed container for a few days until the family went to town. They would take their cream to the "Cream Man" who tested it for butterfat content and paid them for how many pounds they had. Since cream was often accepted only on certain days that would naturally be the day the farmers would come to town. This spawned the ever popular saying that this or that event "had more excitement than cream day in Altonah" or Neola or wherever.

A little later for those with more cream there was a cream man that would come by the farm once or twice a week to pick up the cream. From the cream, butter was churned and either shipped out or molded into one pound packages for distribution locally.

The days of the cream check were like a cocoon from which emerged the milk shipping dairy enterprise with the advent of cheese plants and larger dairy processing plants. The small milking barn with its row of about eight stanchions with a grain feeding manger and concrete floors was born. The Moon Lake Electrical Cooperative brought power to many of the farms and milking machines came into vogue. This made it possible to milk more cows and the numbers increased to thirty or more. The milk was filtered into ten-gallon milk cans and placed in a cold water bath until the milk truck came by to pick them up.

This was the mode when I arrived back in the Basin to practice, and in a short time artificial insemination became a possibility. With the development of the rural telephone system this method of breeding became more popular. My service in checking cows for pregnancy gradually increased, although the value derived was minimal because they would only check cows that had been milking a long time in which they couldn't bump a calf. It was a very difficult educational process to get them to keep good breeding records and check the cows in early pregnancy when treatment might be given to help settle the cows. The fire engine nature of my practice over such a large area didn't lend itself to herd health programs either.

Mastitis, almost unheard of before the milking machine, began to take its toll and improved breeding for increased milk production magnified the number of milk fever cases. The drug companies were full speed ahead in competition to produce udder infusion tubes and drugs to cure mastitis and a new one showed up every few months. It was hard to keep up with the so called "latest thing." The sulfa drugs were coming forth in about six different varieties and penicillin and terramycin were the hope of everyone. The misconception that you could treat yourself out of a mastitis problem was born and still remains alive and well today, even though experience has proven it wrong many times. The cheese plants began to have trouble making cheese and the dairy processing industry and regulatory agencies embarked on a quality control program. Although all milk sold was pasteurized, the wholesome nutritious image was being tainted. Paralleling this, the elimination of tuberculosis and brucellosis in dairy animals was becoming a must. The wings of the dairy butterfly were molting again and out of the next cocoon came requirements for refrigerated rapid cooling and storage, in-line milking machines where the milk went directly to the refrigerated tanks, barns that were clean and painted, water that was potable and stringent limits on bacterial numbers allowed in milk. It required the clipping of the hair around the udder, washing the teats before milking and a much more intense sanitation program.

The investment required for new milking barns or remodeling of existing ones to meet the standards began to separate the farmers from the dairymen. The man who worked at a job and together with his family ran the small dairy herd on the side had to get out. It was the demise of many small dairy operations and required those who stayed in the dairy business to milk more cows to make the payments. Almost simultaneous with conversion to in-line milkers and tanks, came the new color of the butterfly in the form of walk-through milking parlors with lowered milking areas for the operator and stalls along the sides where the cows were elevated enough to do the milking without stooping. This speeded up the milking and made even larger herds possible.

Ioka Dairy

There also grew out of the walk-through barn a cooperative effort where two or three dairymen would build the facilities together and each use the barn to milk his cows in or pay a milker to do it. As you can imagine this was a time bomb for trouble because of the records and diverse operational opinions. One such milking facility was the Ioka Dairy. It started out as a shining example but soon ended up as a one-man operation. The first fly in the ointment I discovered was the fences. The majority owner thought four-and-a-half-foot fences were all that was needed. They looked so nice and you could see into all the pens. So all the posts were sawed off even with the top board. Some cows were a little more athletic than others, especially when they got excited, and they just sailed over that top board. Or they tried to and broke it in the process. Herds got mixed, tempers flared and some cows couldn't be handled as desired. Since there was no other facility for examining or treating a cow short of a long rope, the barn was used for this like most all others were. I guess my activities trying to get the cows in the barn, prompted as much fence jumping as any other.

This dairy enterprise had its share of problems but I knew it was doomed to failure one morning when I stopped there. They were having some mastitis problems and I was trying to help. I was returning from a call up at Altamont. It was about eleven A.M. I noticed Mr. Odekirk's vehicle there so I thought I'd stop and see how the mastitis was. When I stepped into the barn, he was getting things ready to start milking. I thought he must have been broken down or the power had been off or something. I asked him if he had had some trouble. He said, "No. I just decided that from now on I am going to get eight hours sleep and the cows can wait. I had a meeting that kept me up and I didn't get home until one A.M., so I just set the alarm for eight hours."

This brief chronology of dairy farming in the Basin covers an epoch in time in which I was involved, but much of the detail has been left out. As a veterinarian I felt the responsibility to be of service and help the dairymen in every way possible. There was a multitude of information coming forth in all facets of the industry and it was a real challenge to try and keep up to date. There were opportunists who saw a chance to make a few bucks and I had to be able to sort out the good from the not so good. My advice could affect the size of the milk check and it had to be tempered so that it wouldn't criticize or offend. I was constantly working with them on their nutritional formulations. With the production increases acquired through breeding efforts it became more difficult to meet the demands of the cow for reproduction at the same time she was milking heavy. It was very important to keep the proteins and carbohydrates adequate, the mineral requirements in their proper ratio and furnish needed and usable vitamins.

Mastitis

As I mentioned, mastitis evolved as the dairyman's number

one curse and we all had to learn together that milking techniques played the greatest roll in its control. There were differences in the various milking machines and the vacuum levels could vary if machines were not kept clean and in good repair. Different sizes and designs of teat inflations played a roll. Washing, dipping of milker claws and teats, drying teats off in freezing weather, proper bedding and housing all had a part in mastitis control. Proper treatment and follow up saved or lost many cows and isolation and milking order reduced the spread.

I read the journals, attended seminars, kept in close touch with the Colleges and the Extension Service and consulted with other veterinarians in an effort to be informed. Somehow it was against my nature to charge anything for consultation or advice. I just wanted to help them.

Calving Interval

Another big problem in dairy cows was getting them to breed back in a reasonable time. When they are producing heavily you about have to keep on top of the nutrition and management to keep the calving interval where it should be. In many herds, my services were used to find when the cow was going to calve so they would know when to dry them up. This was all right, but it was never as accurate as good breeding records confirmed by pregnancy testing. I think my greatest challenge was getting dairymen to put forth the effort needed to keep good records.

The Challenge of Newborns

Loss of the newborn calves was an economic drain that was many times not recognized for its true value. I think the beef cattle operators were more aware than the dairymen because it was the product they sold in the fall. The dairymen were increasing their herd numbers and getting more calves, but they were not providing the care the calves needed. Typically they just added more calves to the existing pens.

With the increased numbers housed together, the pathogenic organism levels also increased and as they passed through one calf after another their pathogenicity, or ability to cause sickness, increased also. With older and younger calves mixed together in greater numbers it became a vicious cycle.

The newborn calf at birth is not fully equipped to produce antibodies for its own defense. This takes a month or two to develop and the animal is vulnerable during this time. Mother nature provides them with a goodly supply of antibodies in the colostrum if the calf gets a good supply in the first twenty-four hours. These antibodies, absorbed into the blood from the gut, provide a measure of protection if the mother has been exposed to the organisms, but it is a one-time amount. If the calf is severely challenged the supply is depleted and the calf gets sick.

The number one cause of loss was diarrhea or calf scours which stemmed from a multitude of factors. Sometimes overfeeding triggered it, but its basic cause was a combination of pathogenic viruses and bacteria, combined with stress of one kind or another. Weather played a key role by chilling, wetting, and contaminating the calves when resistance was limited. Close proximity led to easier transmission of organisms from one animal to another, and moisture facilitated the contagion. The multitude of causative factors produced the scours, and once the calf began the diarrhea there was a rapid loss of body fluid and electrolytes. This combination of dehydration and electrolyte imbalance soon brought on fatal consequences.

Without proper knowledge and particularly without adequate facilities the calves were crowded together in pens. To compound the problem, newborns and younger calves were mixed with older ones who still carried the diseases, but were producing antibodies to prevent sickness. My challenge was to convince the owners that they must keep each newborn isolated in a clean pen by itself for the first three to four weeks, then put only two or three together for the next few weeks and to keep the older calves entirely away from the younger ones.

With the beef calves I always encouraged the owners to keep the cattle scattered as much as possible, and not bring them into contaminated corrals or feeding areas. If an outbreak did occur and got severe I'd recommend separating and moving the cows that hadn't calved to a new clean area.

Treatments were as numerous as drug companies and home remedies, but success in large measure came from fluid and electrolyte replacement and medicines to control the organisms.

I had many experiences also where calves were grafted onto nurse cows that would raise several calves in a season. Sometimes an older calf would be grafted on with the mother's newborn and the newborn would immediately die. The other scenario was to bring in newborn calves to add to older ones already nursing, and these would die.

Another practice with severe problems was the importation of baby calves by the truck load to be raised as replacement heifers. These heifer calves were gathered and purchased from the milk shed areas of California and Wisconsin for a song and a dance so to speak and then shipped up here to be raised. There were also many purchased from calf peddlers who brought them in. They had two strikes against them when they arrived having been exposed to many organisms and having been stressed in the shipment. There were many that I could not save and all seemed to lack resistance. Even after surviving for several weeks they would get sick and we couldn't turn them around. I had a reoccurring question go through my mind. Why don't these calves develop immunity? Was there something interfering with these calves developing antibodies? As I thought more about it, my flying experience led me to think about red blood cell needs at different altitudes. Out of this I formulated a theory that I feel accounts for the problem. These calves born at sea level don't have as many red blood cells as they need to survive at this altitude. Therefore, with an already immature blood-building capacity their resources were pressed into the production of red cells which were their greatest immediate need. With this demand for red cells, the antibody building function was delayed for weeks beyond the normal. The higher the altitude the greater the problem. I was called upon to look at a herd of calves and necropsy some dead ones for Mr. George Carroll whose place is about six thousand feet above sea level. These calves had been a little older when they came in the spring and they'd survived the summer, although they hadn't thrived and many had been sick. The demand on their hearts to try and compensate for the lack of red blood cells had taxed them to the point of failure. The heart muscle was flabby and enlarged to the point the valves were leaking. The calves were slowly dying of heart failure. He suffered a great loss because most of them died.

Remedies

As I worked through the years, I encountered many old-time remedies for calf scours. One of these was the addition of a raw egg to the milk. Knowing the excellent source of nutrients in the yolk of an egg I concurred in this practice. In fact as the research on electrolyte losses from scours became known, I felt that the added egg could have replaced some of those nutrients. At any rate, it was a widely used practice and like many other facets of livestock management, there were always some who aspired to the philosophy that if a little was good then a lot was better. Arnold Powell was one who was quite successful in raising calves and at the time was driving truck for the Draper Poultry Cooperative. He had access to and obtained a lot of checked eggs that couldn't be sold, and fed them to his calves. At first the calves thrived and did very well. Then, at about two months of age, they began to get sick and die. I was called in and performed some necropsies from which I discovered that the calves had died from kidney failure. With some assistance from a pathological laboratory we found that the kidneys were plugged with albumin. Too much egg white had caused the problem. I learned a great lesson and from that time have always recommended the removal of the white of the egg and the feeding of only the yolk.

The final problem I'll mention with raising dairy calves was grain founder. The calves were usually offered some kind of concentrate or grain on a free choice basis as soon as they began to nibble at solid feed. This worked out fine in the beginning, but as they got older, from two months up, they would overeat on the grain and get sick. They would get wobbly on their feet and then go down, usually having a profuse diarrhea, grayish in color. Sometimes this would happen even when measured quantities were fed and one piggish calf would get more than its share.

Other disease problems came along and it was always my thinking that immunizations and other preventive measures were more important than trying to treat animals after they were sick. It was a privilege to work with many wonderful families and feel that I had contributed in a positive way.

Calf Diphtheria

I often got calls from owners about calves with severe pneumonia. At least to them it appeared to be pneumonia because the calf would be breathing so hard. Their description of how hard the calf was breathing, especially if they described a loud rasping noise, alerted me to the real problem. Diphtheria was an infection of the vocal cord area of the larynx with an organism called spherophorus necropherous. It caused a lot of swelling and some of the tissue to die, narrowing the air passageway until breathing was very difficult. If animals were treated early, with the proper medication, recovery was possible, but sometimes more drastic measures were needed.

I was called out one day to a ranch and among other problems was a calf gasping and struggling to get enough air. When we began to treat it, the extra exertion was too much and it passed out and stopped breathing. I knew its only hope was an emergency tracheotomy, so I pulled out my pocket knife and made an incision over the trachea under the neck. Then I cut between two tracheal rings to make an opening. With my finger under the trachea to hold it up and the hole open, I put my mouth over the opening and blew air down into its lungs. With several more forced breaths it began to come to and breathe on its own. It was a great feeling to see it start to breathe and begin to respond. However, when I would let the trachea back into place it closed off and the calf couldn't breathe. I needed a trachea tube to place in the hole to keep it open and I didn't have any. In fact there were no trachea tubes made that were small enough for calves. What was I going to do? Then all at once the light came on in my head. I pulled out my ball point pen and took it apart. By cutting off the tip I had a plastic tube about the right size and when I slipped it into the trachea the calf breathed fine. The only problem was that it wouldn't stay in place. To solve this I drilled two small holes with my knife and ran some suture wire through them and stitched it to the skin. It worked fine and the calf soon got up and we finished the treatment. In a few days when the calf was able to breathe on its own, I went back and removed the tube. It took a little ingenuity but it worked that time and a good many more times over the years. It was the fulfillment of the old adage that necessity is the mother of invention.

Grain Founder in Cows

The curiosity and uncanny abilities of some cows often lead them into trouble. They would unlatch a gate, open a door, or pull a sack of grain out of a truck. Sometimes they would keep licking at a place where a small opening in a grain bin was leaking. They would eat until their stomachs were overloaded and it would become toxic to them. Owners were well aware of the dangers, putting two locks on the critical gates and instituting other safety measures, but invariably some old cow would get into trouble. When the problem was discovered or when the cow began to show symptoms they would call me.

They told us in vet school that you could "lavage" or wash

out a cow's stomachs. However, I was never privileged to see it done. When my first case came along it was about the only option short of surgery on the stomach. The owner and I put the cow's head in a stanchion, then put some nose tongs in her nose, pulled her head around and tied it down as near to the floor as possible. I pushed the large Kingman stomach tube down her throat and into her stomach. With it in place, I stuck the water hose into the end of it and we ran water into the cow. When the cow was filled and distended to the point of discomfort, I would pull out the water hose and hold the stomach tube on the floor. It was amazing how that water and digestive juices siphoned back out of that stomach and brought large quantities of the grain with it. I'd have to jiggle the tube in and out to keep the wall of the stomach from sucking into the end and stopping the flow. When the cow looked emptied out and the tube stopped flowing, we hooked up the water hose again and filled her back up. By the time this procedure had been repeated four or five times most of the fermented liquid and a goodly portion of the grain was on the barn floor. The poor old cow was thoroughly chilled to the point of shaking from having all that cold water run into her, but she would now live.

Great Ladies

"Dr. Dennis, this is Rita Hansen. Our milker tells me we have a cow with a bad case of mastitis. He thinks she might die. He said he'd treated her, but it didn't help much. Have you got any secret medicine that works? We must have some new bug or something. Anyway we seem to be having more trouble than usual; can you come up?" People often worried about new organisms. Grant and Rita Hansen had a sizable dairy operation about halfway between Upalco and Mt. Emmons. It was located on the brink of the hill about a mile west of the waterfall. I'd been to their place many times to dehorn, castrate, vaccinate and treat many of their animals. I thought about their location and remembered when I was growing up how adventuresome that old road was. It went west along the hill from Upalco, then snaked around a bend and crossed over the canal right at the foot of the waterfall coming off the end of that bench. After crossing below the falls it turned west and climbed to the top of the bench where you could look in all directions at a vast panorama of the Basin. I remembered how awe inspired and excited we would be, because if we ever went up that way it was to a family camping and fishing trip. The road had its share of rocks and we weaved and bounced, but oh, what fun it was. The old timers always built their roads along the south slopes where the snow melted first and the gravel and shale dried fast after a rain. This road ran right along the brink of the hill for about four miles where it would stay dry and the ever present west wind would sweep the snow away as it lashed at the edge of the bench. When the new road was made and paved it took a more direct route up through the pastures and the thrill and spray of the waterfall passed into history. The well-drained edge of that bench with the ever flowing canal was an ideal location for Grant and Rita's farmstead and they had built some good corrals and silage bunks. The milking barn sat right on the edge of the hill.

"Yes, I can come. Would it be all right to come about milking time so I could check things out with him?" Mastitis was an ever present threat with many contributing causes and I wanted the man doing the milking there for consultation and directions. From the sound of things there were more than the one sick. Trying to separate them from the herd during the middle of the day could be very time consuming. There were two big corrals with upwards of a hundred cows in each and there wasn't an alley way with gates to separate them and get them to the squeeze chute. Sometimes I had roped them out in the corral to treat for various things, but it stirred up a milking herd.

"Whatever is best for you. He should be here about four thirty." Rita was always very cooperative and cheerful. I had never seen her upset and admired her willingness and ability to help with the farm.

"What time will Grant be home?" Years before, Grant had lost an arm in a corn chopper accident and had acquired a contractor's license where he worked quite a lot of the time.

"He won't be home till the weekend; maybe not then. He's building a school out by Ogden and he has to stay out there most of the time." There she was with the whole responsibility of the dairy farm and not a hint of complaint in her voice.

"Oh, you've got it all to yourself. You can do as you please. How lucky can you get?" I joked with her knowing that she and Grant shared in everything they did.

"Well, it's a little better this time. Where Grant's going to be gone so much he hired this full-time milker, so I don't have to do any of it any more." I could tell she appreciated the arrangement.

"That sounds like a good deal. I'll be there about milking time."

I wasn't able to get away right when I should have, so I was an hour late when I pulled up beside the barn. The milking machine was purring as I got out of my truck and entered the barn. When I paused in the door to the milk parlor I could hear the pulsating milking claws as they alternately sucked and relaxed on the cows' teats. The hired milker didn't seem to be anywhere around and there were four cows with udders long since drained that had the milkers still on them. As the milkers sucked away at the udders the teats were being pulled deep into the inflation cups. I knew they should be taken off, but I didn't want to interfere. With the development of the mechanical milking machines, mastitis had mushroomed from a rare occurrence to a major problem and damage to the teat from milkers being left on too long was the primary cause. How could I impress upon this guy that milkers shouldn't be left on after the milk was all gone? When I looked out the partially opened door to the catch pen I could see and hear him trying to push the second herd in and close the gate. I put things together in my head and realized he had left those milkers on while he went after the other herd of cows. That could have been fifteen or twenty minutes. I could understand why he was having mastitis problems if he didn't pay more attention to getting milkers off than that. I started at the front and took the four milkers off and hung them up. He came through the door after checking the tank and looked to where I had removed the milkers.

"Gee, thanks. I guess you're Dr. Dennis?" he said as he went over and opened the gates so the four cows could go out.

"Yes. I'm sorry I'm late. I couldn't get away. Is that sick cow in this herd or the other one?" I was feeling guilty about not getting there before the milking started.

"Oh, it's okay. She's in this bunch. When you weren't here I milked the other herd first. Have you got any extra penicillin with you? I'm about to run out. I've sure had to pump a lot of it into 'em. There must be a half dozen cows with mastitis in each herd. If you'll stand here in the door I'll go bring the sick one in. She's right in the back. I'll bring her up around this side." He was pointing to the south side.

"I'll do it," I said as I stepped over a ten cc metal syringe with an udder infusion cannula on it. It didn't look like it had been washed for a week. He went down around the herd of cows clearing a partial path as he went and tapped the sick one to start her up through. She reluctantly staggered up the slope toward the door as her swollen udder was pushed from one side to the other by her legs. I stepped out to hold the other cows back and she lumbered forward to the first stall. I followed, closing the stall gate and took her temperature. It was a hundred five and a half. When I went down and began to milk her the stuff came out in huge curdled lumps.

"Her milk got so bad I just quit milking her," the man said as he took a peek and went back to letting cows in and putting milkers on them. I kept milking until no more would come out and washed the slimy puddle down the drain under the cow. The cow acted much relieved and I went to my truck for an intravenous treatment, a box of infusion tubes, a bottle of combiotic and a syringe. When I finished treating the cow I explained that she needed to be milked out several times a day if possible.

"How have you been treating the cows with mastitis?" I asked as I followed him to a cow.

"I just shoot this gun full of penicillin up in the infected teats after I milk 'em," he said as he picked up the dirty syringe. I knew he meant combiotic when he said penicillin. We all referred to it that way.

"Do you follow up the next day?"

"Only if they show lumpy milk. When I wash and check them, if they got any flakes or lumps in the milk I give them some more." It seemed his treatment regimen was check the milk and shoot the medicine in the teats that showed trouble.

"Do you mark them some way so you can keep the milk out of the tank?" I had a horrible feeling that a lot of milk was going into that tank that shouldn't.

"Naw. I remember 'em and milk them in this other milker where I can feed it to the calves if the milk isn't good. It sure is a lot of bother changing those hoses back and forth and I'm getting more now than I need for the calves." I concluded that the only way he remembered was by checking a squirt of milk.

"Have you ever thought about milking them last after all the others are done?" I could see that milkers were going from infected cows onto healthy ones and he wasn't practicing any dipping routine between cows. He did wash one milking claw off with the hose after it fell on the floor.

"Are you crazy? There ain't no way I could do that. I'm lucky to just get 'em milked and the calves fed." I was getting more discouraged by the minute, my stomach was churning and I didn't know what to do. About that time he detected another infected cow and switched the milker hose. When she was done I suggested we use the tubes of mastitis medicine I'd brought in. I showed him how to wash the end of the teat and put in the tube without contamination. I explained that he could give an intramuscular shot of the combiotic if there was heat in the udder in addition to the tubes. The dairy industry, drug companies, and even the veterinary profession were still trying to treat themselves out of mastitis trouble. There wasn't any testing for adulterants in the milk at that time and keeping antibiotics out of the tank was the responsibility of the dairyman, most of whom tried very hard. How was I going to get through to this guy? I'd brought a marking crayon in and I made a mark on the cow so he could tell her apart from the others.

"Now you need to treat her for at least three days even though the milk clears up, and keep the milk out for at least two more days after the last treatment." He mumbled something about pouring milk down the drain and I could see by his expression that his mind wasn't changed. I stayed and treated and marked several more cows and brought in all the supplies I had in the truck.

When yet another cow showed up with mastitis he said, "Wouldn't it be a lot quicker to give her a gun full of penicillin like I've been doing? I've tried those tubes and they don't work as good as the combiotic, I can tell you that." Sanitation and prevention of spread from cow to cow hadn't seemed to enter this guy's head and getting the milkers off when the cow was done only happened when his routine coincided with it.

I could see a disaster in the making, in fact it was already there. I gathered up my nose tongs, empty bottles, IV set and syringes, and went out to the truck. It was nine o'clock, but the lights were still on down to the house. I had to talk to Rita about what was happening. Maybe she could talk to the guy. I knew she had milked a lot and knew what should be done. These were the tough times in practice, when you had to try and tell them bad news without painting a bad picture of somebody. When I knocked she invited me in and introduced me to a couple of the kids that were studying. I think she always wore a smile or had one in waiting and was a delightful and pleasant person to be around.

"Well, how's the cow?" she asked without a hint of apprehension in her voice. I still hadn't decided how I was going to approach the problem.

"She was pretty sick and I think her udder is shot, but I think she'll make it if she's milked out. He hadn't been milking her out."

"He wasn't milking her?" She made a mental note. "What about follow up treatment? Will she need shots?"

"Yes. I told him she would need about twenty-five cc's of combiotic a day for three days and be milked out three or four times a day if you have someone to do it." I think she could read that I didn't have much confidence that he would do it.

"What do you think of that milker we've got?" There it was point blank and yet like a pleasant inquiry. I knew I had to spill the whole story, but there was one piece of the puzzle I didn't know.

"How long has he been doing the milking?" I was thinking about what part to lay out first.

"He's been at it three weeks. Why?" I guess she wondered what that had to do with it.

"Rita, I don't know how to tell you this, but I don't think he's had much dairy experience. I think your cows are in trouble. When I pulled up tonight and went in the barn there were four cows with the milkers on and he was over getting the other herd. It doesn't seem to matter to him when the milkers should come off. It's a matter of convenience in his milking routine. As for mastitis, when he washes the udders, and that was one thing he did fairly well, he squirts some milk out of each teat and if they show anything he treats them *after* they're milked. He doesn't mark them and he takes the contaminated milker and puts it on the next cow. His treatment consists of a gun full of combiotic in the teat with a syringe that doesn't look like it's been washed since he came. I'm sorry. I shouldn't let my frustrations show so much, but his only concern with mastitis is check the milk and pump in the combiotic." I stopped and waited.

"I appreciate your frankness, Dr. Dennis. I've had my suspicions. There have been several cows go to the sale and our milk volume is down considerably. I've hated to look down his neck, since Grant hired him and got him started. Men usually don't like to be bossed by women either. I'll have to call Grant and we'll do something. You told him what to do with the sick one?"

"I tried, and I tried to get him to milk the infected ones last. I told how he should follow up to get them cleared up and should mark them and keep the milk out of the tank. I don't think he wants to know what to do. Leaving those milkers on too long is going to get a lot more cows in trouble." I had told it like it was.

"I appreciate you telling me. I hate to hear it, but at least I know my suspicions were right. I'll bet you're tired." I'd moved toward the door with my hat in my hand.

"Something like this tends to make a guy tired, but I'll get over it. I guess I'd better head for home."

"Thanks, Dr. Dan. If I need you, I'll call. Be careful going home. Don't fall asleep." I hurried to the truck and was on my way.

Sometime during the next week I was up that way about milking time and drove up to the barn. There was Rita flitting back and forth milking those cows and probably singing if I could have heard her. She had a smile and looked as lovely and well groomed there in the barn as if she was downtown.

"So are you the milk maid tonight?" I asked as I paused in the door.

"Yes, and in the mornings too. I canned that bird the next morning after you were here and decided to do it myself." I thought what a dedicated and willing worker she was. After a few questions and answers I left her some more medicine and headed for home.

As I drove along I thought how families work together in the dairy business, and how many other wives and children I had seen helping out. When I got to Upalco, another person filled my memory. It was Bernice Mitchell another gracious and lovely dedicated wife—a lot like Rita. Her husband would have to go on construction and she would take over the dairy herd. When there was trouble she would call me and her only negative comment would be, "It seemed like the cows always wait until Mar D is gone to get sick." She was always cheerful and happy in a quiet sort of way and as lovely milking cows as in her home. I thought two finer companions to their husbands have never crossed my path. I've always admired them both.

They Don't Stay Milked

I could tell that the trying obligation of having to be there to milk twice a day weighed a little heavy at times on some dairymen. One experience I'll not forget was at Med Benson's one evening. He had called me out to remove a retained placenta from a cow. Med was getting along in years, probably in his eighties. When I arrived he was just finishing the milking of the last two cows. The cow I was to work on was in the end stanchion. When he had dumped the milk and set the milkers away he sat down on a small bench at the back of the barn like he was tired. As I stepped by him I asked him if he was a little tired. I've always remembered what he said.

"You know, there's only one thing I don't like about milking cows."

"What's that?" I asked turning back to look at him.

"They just don't stay milked."

I think that was a feeling that most dairymen shared at one time or another in their lives.

I remember the wisdom of James Hamblin speaking to a group at church one evening. He was telling all about his life; of coming to the Basin and milking cows for a living. He told how he had enjoyed working with the soil and the livestock, especially milking cows. Then he said the thing that he appreciated the very most of all was the oil well that had been drilled on his farm. I think most dairymen would feel the same.

11
CONDITIONS RELATED TO PREGNANCY

Irv's Preg Tests

Being soiled, dirty, and a bit smelly seemed to go with the territory in practicing veterinary medicine. Even when working on the front end of a cow the cough or forceful expulsion of nasal mucus found its mark, but the other end was something else. In the morning I had been out delivering a breech calf from a big Holstein dairy cow. Having to reach into the very last millimeter of my arm, the green fecal flood flowed right up my sleeve, but that old sister wasn't content with that. Every time I started to manipulate the calf, she would strain, forcing placental fluids out and then releasing a jet of urine behind it to shoot up my arm.

Having completed the job, I flushed off enough to keep it

off my truck seat and headed home. At the office I peeled the outer layers off with my coveralls, then went on home to shower. Once free of my soiled clothes, a generous application of soap and brush took away everything but the smell on my hands. This I either removed or masked, I've never been sure which, with applications of Clorox bleach. At any rate, it felt good to be spic and span again in some clean-smelling clothes, and to have my lunch. My wife was a talented angel that could feed me at any hour and get that muck out of my clothes.

That afternoon with nothing on the agenda, it was a rare pleasure to sit there at the desk all cleaned up and do book work. I'd even washed my boots and oiled them.

"Who was that?" I'd turned in time to see the back end of a pickup with cows in it go by the window.

"I don't know," Nancy said. "I didn't get a good look at the driver, but he did have two cows in the back."

"Anybody call you about coming in?"

"Nope, there haven't been any calls like that." About that time Irvin Secrest came around the corner and walked up to the counter. Irvin was a round-faced jolly type of guy with a small farm out in the Ballard area. He had a few head of cattle and when I'd gone to his place his corrals and facilities were limited and often in need of repair. I knew he was struggling financially, working most of the time, and the cows were kind of something on the side, like a lot of other farmers. He seemed prone to not do anymore with them than he had to, like they were a burden to him. At least that was my impression.

"Hi, Irv, what can we do for you?"

"I got a couple a cows to preg test."

"Okay, I'll be right with you." I put on my hat and went out into the back room where I pulled on a nice clean pair of coveralls. When I stepped out the back door, I could see that he hadn't backed up to the unloading chute. "Why don't you back up there to the chute, and we'll unload those old gals and put them down the alley?"

"Well, Doc, I was wondering if you could do 'em in the truck. They're buggers to load." I'd been down this road many times before, and it seldom turned out very pleasant. But for Irv, maybe I could do it. I stuffed the sleeve in my pocket and looked up at the wooden homemade rack. It was wired together at the back, and literally dripping with liquid manure, I wasn't going to need any lube. I had to get my hands dirty anyway, holding the tail, so I grabbed on and climbed up keeping my coveralls out away from the rack. There wasn't much room for me in behind the cows, so I reached over and tapped them on the tailhead to move them forward as far as they could go. The top board was high enough that it had escaped most of the green so I put my leg over and eased myself down with one hand on the tailhead of a cow, and the other on the rack. That tailhead was the only dry place on either cow. I thought, they sure must be loaded with fluke to be that loose. I'd made it down to the floor of the truck without being hammered by one of those hind feet. In that close of quarters, you can't escape those sledge hammer blows, so I kept my shins turned sideways a little and hoped I wouldn't get a broken leg. I wiped my left hand on the top of the cow and pulled on my plastic sleeve. Touching the cows and talking to them, kept them up front. Now, trying not to touch the back of the rack, I took hold of the tail of the cow on the left and lifting it, I rubbed my hand around to take advantage of the green lube. As I pushed my hand into the rectum, the cow leaned forward at first then relaxed. I cupped a little of the liquid manure out on to my glove then began to slide my hand forward to the cervix and uterus. Her patience evaporated and she lunged forward trying to go around the front of the other cow. Naturally, her feet slipped in that inch-deep layer on the truck floor, and she splattered me up to the knees. The other cow responded by backing up and trying to turn, and this pinned me right in the corner. I could feel the coolness of the liquid manure seeping in from the truck rack on one side and the warm manure she was running down my side and into my pocket on the other. One foot had scraped down my leg to my boot top and then landed on my toes. I dug my elbows into her back and hip and heaved her forward using the truck rack behind me to get her off my foot. I knew I shouldn't have got in that truck, but it was too late now. I grabbed that cow's tail and shoved my hand in to my elbow and felt a calf's head. With that and a slap on the back, she pushed the other cow back around and I did what I had to do to her. My toes hurt a little as I climbed back out of the truck, but she hadn't put much weight on them. I began to scrape the worst of the manure off with my hands and throw it on the ground. Irv was as unconcerned as if that was routine, and I started toward the door.

"By the way, Doc, take a look at this one cow's eye," he said pointing to the one that had pinned me in the corner. I walked around the truck and got a look at the eye.

"That's cancer on the third eyelid," I said.

"Can it be fixed?" He was looking up at the cow.

"Yes, all you have to do is trim off that third eyelid, and the cancer comes with it, but we'd have to have her in a squeeze chute to hold her head still."

"That's no problem. I can back up to that chute and have her out in a minute."

Vaginal Prolapse

A vaginal prolapse occurs when internal pressure bulges up the floor of the vagina together with the bladder, and they are forced out through the vulva into a round balloon, sometimes as large as a basketball. It happens when the mother cow is heavy with calf, and within about ten days to two weeks of delivery. What stimulates them to do it is a real enigma. We speculate that the internal abdominal pressure of late pregnancy plays a part, especially when the cow is lying down, but what stimulates them to strain has been attributed to many things, and no one really knows. This has been one of those times when I wished I could talk to the cow and ask her. Some speculate that it's hormonal, others think it might be mineral imbalance in the body. Most think fecal contamination of the mucosa causes irritation. No one really knows, but the cow is found with this balloon hanging out the back end and straining as hard as one in labor. The bladder is pushed out into the balloon, and urination is very difficult. If not treated early, the mucosa thickens, dries out and hardens, and the back pressure on the kidneys compromise their function. If left too long, adhesions form inside the prolapse. With the cow generating all the pressure she could to push it out, it was no small chore to replace one, even in a squeeze chute. Once replaced to normal positioning and sutured to keep it there until the birth occurs, all is well until the next pregnancy.

When a cow once had a vaginal prolapse, they nearly always repeated it the following year, so I always recommended that they be sold for slaughter when the calf was raised. Not everyone followed my advice, especially when the calf was born and no more trouble showed. I remember one man who made it a point to tell me I was wrong when his cow didn't repeat the next year. It seemed like they enjoyed pointing out my mistakes, especially if there were other ears to hear it. I did appreciate his honest character, however, when on the following year he came in and sheepishly told me she had prolapsed and he'd found her dead.

The office and small animal facilities in our basement worked out very well, but the doctoring of the large animals in the trucks in the backyard left a lot to be desired. I recall the vaginal prolapses in particular. I would get up into a pickup with the cow's head tied as close as possible up front, but it didn't leave you any room to step back and escape those sledge hammer legs as they kicked. I'd get my suture material all ready and have it in my pocket before I washed them off and pushed them back in. If I could get them all cleaned up first, then lift them up to relieve the urine flow and squeeze out most of the urine, the cow would relax her pushing just a little for a second or so. When she did, I'd start with the bottom and push for all I was worth, and most of them popped back inside. The easiest way to keep it there was to push my left arm and fist up inside while I started suturing.

I received some good advice on suturing from a Wyoming

vet at one of our Intermountain Veterinary Medical Assn. meetings. He said, "Go out into the haired area where you've got some good rawhide to hook into, then cinch her up so tight she can't get her eyes shut." It was good advice, but when you start jabbing suture needless through the skin, those cows are not very happy about it, and they kick viciously. This was the challenge then, trying to stand back on the endgate of a pickup with your feet, and lean forward holding in the prolapse with your left hand, while you thrust the needle through the skin with your right hand. I learned that if you could stand thin enough exactly in the middle, most of the kicks missed your shins and knees, because the cow has a tendency to kick a little bit laterally. The other thing that worked sometimes was to shift to the left as you stabbed the needle in on the right side, then over to the right for the left side. The only trouble with the pickup was that there wasn't enough room when you pulled out your hand and hauled back on the suture tape hurting on both sides. The cow could dance back and forth and kick faster than you could move, so you couldn't entirely escape. I came to the conclusion that a squeeze chute, and a place to use it, would be a great help. It was ten years before I could afford to get one.

The vaginal prolapse is always a bit of a challenge, but sometimes when both the rectum and vagina are expelled, they are quite formidable. One experience I remember occurred after I had my chute, corrals and animal hospital. I had a cow in the chute. She was a big cow and had two very large balloons pushed behind. I was washing and cleaning her, when a drug salesman stopped. He came out to the corral to see me. There was an old fellow who used to come down and help me, now and then, standing over by the fence. The salesman watched a minute, then asked him, "He can't hope to get all that back into that cow, can he?"

"Oh no," Mr. Shiflet said. "He just stretches her hide out over it and sews it up."

I think he must have believed him because he just said, "I'll see you another time," as he spun around and left.

Uterine Prolapses

I mentioned that vaginal prolapses could be a challenge and that being the case, complete uterine prolapses were extremely challenging and critical for the cow. With the complete uterus turned wrong side out, and sometimes some of the intestines pushed out into it, it could be a washtub full of trouble. Most cows with a complete prolapse are very sick, usually in some shock and reluctant to stand. The more wild cows sometimes jumped up and tried to run with all that hanging on the back of them. With all that weight hanging, it sometimes ruptured some of the uterine vessels, and they would hemorrhage to death. If a cow was caught and got on her feet, I'd rush up to her and hold up on the uterus to try and prevent the vessels from rupturing. We always tried to keep the cow down, if we could, because it was easier and safer.

The replacement of a uterine prolapse is a long, tiring struggle and usually requires some help. There were times when I found it impossible to out push the cow and had to resort to hanging them part way up by their hind legs. From the experience of that first one I had in Idaho, I learned that if the cow is rolled up on her brisket, with hind feet pulled out behind her, it's easier to work the uterus in, than if they are laying flat on their side. One way you are pushing partly downhill while the other way it's partly uphill. I also learned that a smooth flat surface to lay the uterus on makes it much easier to clean, and keep clean, while you're working it in. I used a small porcelain tabletop placed under the uterus and across the hocks of the cow. This made it easier to wash and clean, and held it up while I pushed to get it back in.

Sometimes when the cow insisted on standing up I could, with someone to help, hold up on the uterus, get it cleaned and pushed back in, but it was always a lot more work. The cows have a tendency to hump up and try to get their hind and front feet all together, and that makes things uphill. The most discouraging aspect was when I'd have most of the uterus started in, and my pushing was about on balance with the pressure from the cow. My persistence was just about to win, because the cow had to stop and take a breath, but instead she kind of passed out and flopped over sideways on the ground, throwing the whole thing out and into the manure and dirt. It really taxed my patience, and it took a lot of courage and determination to start over again.

Some of them just insisted on dying, like a wild one I went to see for Clair Oberhansley. He told me to go down along the outside of the fence, and he would rope her from his horse and then bring me the rope, and I could dally her to a post. As he approached her, she jumped to her feet and started off on the run. He spurred his horse after her and threw his loop around her neck. When he stopped his horse, she hit the end of the rope, the uterus just kept on going, and ten or twenty feet of intestine came out too. She flopped down and was dead by the time she hit the ground.

Another case presented me with another shot in the dark, or pioneering experience that paid off. I was called out to see a cow that had been lost in the cedars for a few days. When she was found, she had a uterine prolapse. She was down, weak and really gaunted. The uterus was dark and rotting with several big holes torn in it and very smelly. My first impulse was to recommend putting her out of her misery, but the owner was one of the typically humble, struggling farmers who had faith and confidence that I could help. He didn't say anything as I looked things over. My inner self said, "You've got to try." I went back to my truck for the water, and thought maybe some fluids and dextrose might help, so I got them along with an IV set. When I knelt down behind her and lifted the uterus up, there were maggots crawling in and out of the holes and fly eggs all over it. I thought, "That cow can't live with that back inside of her," so another thought was born. I wondered if I could cut the thing off, maybe I could ligate the big vessels and control hemorrhage, if I could find them. Then my eyes fell on the rubber tubing of the IV set. Maybe if I wrapped and tied that around it high up toward the body, I could cut off the rest and it wouldn't hemorrhage. I explained to the owner that I didn't think she could live if we tried to put it back in, but she might if we could cut it off. I wrapped and pulled each wrap tight around the upper end, then tied a lot of knots in the rubber. I carefully began to cut it off. There wasn't any hemorrhage and the tissue up by the neck looked very much alive. When I got through cutting it off I wondered if the stump would slip back into the vagina and slough off there. Sure enough it popped right up inside and stayed without suturing the opening. When I pushed the stump up in the cow, she decided to get up, and walked away toward the watering hole. I got the report the next week that she seemed as good as new, so another procedure was born, at least for me. I've used it many times since and it has been very successful. Sometimes in pulling calves, I found the uterus torn, so after I got the calf out, I reached in and averted the uterus and amputated it.

Preg Testing

Pregnancy testing by rectal palpation was a very valuable tool and I didn't like to hurry too much, because I felt accuracy was more important than speed. I found that if I went too fast the nerves in my arm would partially paralyze from the pressure and I couldn't feel as accurate. There were a few times when things worked well, and I surprised myself. One was a small herd of ninety-five cows over at Manila. It was the first time for the owner, and he thought we would have to put each one through the chute. He had the cattle in a well-built long alley that wasn't too wide, and I suggested we just work them there. With him in the alley, and me to keep the cows from turning, I could work across the exposed cows on the back end and when they were done we could back them out and send them back into a pen. It was an ideal situation, and we had the whole ninety-five head done in about forty-five minutes. I was amazed how fast it went and he was doubly so.

With the realization and acquired knowledge of your own experiences, you learn some truths that are sometimes hard to convince others of. This was the case with pregnancy testing the Nutter herd. Over the years, I had learned that big, fat cows that had lost a calf in the spring, or hadn't been bred the previous year, would breed up earlier and be farther along in their pregnancies than the average of the herd. There were, of course, the sterile ones mixed in with them. Pregnancy testing wasn't something they did routinely with the Nutter herd, but one fall the foreman convinced Mr. Howard Price that it could save them some feed, and help cull out some of the unprofitable cattle. When I got out to the ranch, they had one pen of thirty-nine head cut out and penned separately. It was their standard procedure to cut out the "fat drys" and sell them in the fall.

As I tested the herd, I kept thinking we should check those that were cut out, because I felt we would find a lot of pregnant cows in them. When we finished the big bunch, I asked them if they didn't want to do the others. They assured me that there wouldn't be any calves in them. I stuck my neck out, and told them I thought there would be more than half of them with calf. I didn't know whether they thought I just wanted to make a few extra dollars or what, but they didn't want to do it. I finally told them if I didn't find enough calves to make it worth their while, I wouldn't charge them for doing it. Reluctantly they got the cows in the alley, and as it turned out there were only six or seven that didn't have big early calves in them. It was a lesson that took some persuasion, but it was valuable in the end.

Then for what I remember, as about all I wanted in one day, was a herd I worked for Zane Christensen up west of LaPoint. They were big, fat cows, a little wilder than most, and the mud in the corral was deep and sticky. There was a small triangular pen leading into a long narrow chute and we couldn't get too many in at a time. We'd slog through the mud and get about fifteen head in the pen then I would get down in there and begin to work them. Part of them I had to do in the narrow alley, and this entailed climbing up the side after every cow. One time when I was testing about halfway up the alley, a snotty, stirred-to-anger cow came up behind me and hit me in the butt with her head. She threw me up in the air, and when I came down, she hit me again, and this time when I came down, I was straddled of her neck and was able to fall over the fence. With the number of times I got kicked, jostled, and squeezed with my feet stuck in the deep mud, it was a long day by the time I finished. When we counted them out of the corral, there were only three hundred and eighty head, but it felt like a thousand to me.

The worst kicking I ever got was from a big, black, bully cow in a makeshift twelve-by-twenty pen over in Sand Pass at the lower end of Pleasant Valley. We were crowding them into one corner and doing what we could, then letting them stir around for some new candidates. We had them all done but two, with one an extra wild cow ending up on the back end this time. I moved forward taking hold of her tail just as she came up with both hind feet and caught me on one hip and in the belly. The force of her kicking picked me up and slammed me across the corral into the fence, where the top pole hit the back of my head. I saw stars and blacked out for a minute or two and had a hard time getting to my feet. When my head cleared, I carefully checked the other cow and was about to give up on the wild one, but she got herself wedged in low between two others, and I was able to palpate her.

Over the years I have pregnancy tested many thousands of cows, but the most I ever did in one day occurred on a snowy day at the Indian corrals at Mt. Emmons. We were testing heifers and had enough that I thought it would take two days. I asked them to be ready by eight-thirty. There was a triangular pen leading into the chute alley, which held about forty-five head. When we started filling the pen, it started to snow rather heavy with big flakes. Thinking it wouldn't last long because of how fast it was falling, I got down in the pen and began to palpate the heifers there. They were wet, and it wasn't long until I was as wet and smeared with manure as the heifers. It made things slick, and although my traction wasn't the best, I could push around through those heifers with ease. I'd work one corner, then the other, and if there were wild ones, we'd get them in the chute alley. When the cowboys were filling up the pen, I would get a short rest, but being all wet I got cold, so it was good to get back to work. The snow just kept coming and so did the heifers. We worked on through the day, not taking any time for lunch. It was late in the afternoon when we finally ran out of heifers, and the snow, by then, had piled up to about fourteen inches. When they finished tallying up their head counts of the heifers they had turned out, and the open ones still penned up, we had tested seven hundred and forty-five head. We were sure glad we didn't have any more to do the next day, because the wind was whistling and the temperature was below zero.

Sympathetic Patrolman

"Hello, Bishop's office." I was at church on a Sunday morning trying to catch up and get ready for priesthood meeting.

"I have an emergency. It's a cow that's prolapsed. Their name is Rasmussen, on the Strawberry River above Duchesne. They said you knew where they lived." It was my wife speaking with a touch of resignation in her voice, and I knew why. We didn't like to work on Sunday, and she would be alone with the kids at church.

"Okay, I'll be home and change my clothes." I tried to hide my frustration this time. I knew that many times I complained to the point she would share in my problems. I explained to my counselors, gave last minute instructions, and hurried home. I was good at changing clothes, especially on Sunday. I always joked that I wore my clothes out from the inside, getting in and out of them. I pulled off my dress boots, shed my pants and shirt and pulled on my Levi's. Then I sat down on the bed, put my boot socks over the ones I had on, and pulled on my work boots. I picked up my short-sleeved shirt and put it on as I headed for the truck. My coat was already behind the seat. It didn't take long.

At the clinic, I hurried and filled up four jugs with warm

water and put them in the truck. I checked my needles, suture, fluids and little two-by-three-foot table top. I was all ready to go except for a bottle or so of penicillin. I shivered a little as I paused for a moment in front of the gas heater blowing warm air. It was March, but that breeze was coming from a glacier someplace. The phone at the clinic rang.

"Hello. This is Dr. Dennis."

"Dr. Dennis, could you come up? We have a heifer with her insides out. I had a hard time pulling her calf and this came out right after I got the calf. The calf was dead." The words were coming out like water over a waterfall.

"Who is this?" I cut in, not wanting to hear any more play-by-play under the circumstances.

"The Harveys, you know, up in Tridell." He hesitated, I guess, to let it sink in.

"All right I've got you pegged, but I have a problem. I've got one with the same thing wrong over in Duchesne. I was just walking out the door. I'll have to go do her first, then come to your place. Leave the heifer alone, and she probably won't try to get up and move around much. If she stays down she may strain some, but she'll be all right until I can get there. Have you got her in someplace?"

"Yes, she's in a pen under the shed in some good clean straw. I'm not sure whether she can get up, she might be paralyzed."

"Sounds good. Just leave her alone, I think she'll be okay until I can get there. It'll probably be about two and a half to three hours." I hung up the phone.

The roads were dry, and I made good time on the way to Duchesne. It was Sunday, and there were few cars on the road. I didn't really want to work on Sunday; it was the Lord's day, but what could I do. I was the only veterinarian, and these kinds of cases couldn't wait, or they would die. I mulled the problem over in my mind as I drove along. If I could get these two done, I could still make the sacrament meeting in the evening. It was the most important anyway. I sure appreciated all that my good wife did to take the kids and to teach and help them with their assignments and talks. A lot of thoughts about our life slipped through my mind as I covered the miles.

When I entered the pen at the Rasmussen's with my jugs and pan, the cow struggled to her feet with the averted uterus hanging heavy from behind. She had a rope on, and the boy dallied her head to a post as I hurried to put my arm under the uterus and lift it up. I'd learned that all that weight pulling down could rupture a vessel inside, and she could hemorrhage to death in a hurry.

"Bring a jug of that water and we'll do her standing up," I shot at them. "You hold her tail and pour that water over it slowly while I hold it up and wash it off." It was caked in mud and manure but it washed away easily except in the crypts of the cotyledons. I turned, squeezed and rubbed until I thought it was clean enough to go back inside. My back was screaming from holding up the weight, and I'd long since learned that the natural flushing action of the postpartum uterus would remove a multitude of foreign matter.

"All right, you hold up on this end, and I'll start working it in. Keep pushing it right up against her," I instructed the father. I started grasping hold of the wall of the uterus and pushing it through the vagina one side and then the other; some on the bottom, and then the top. It was hard work and my arms were feeling the strain and my breath was getting more labored. As soon as you start to push, things inside the cow must feel like she has another calf to get rid of. She was humping up with all four feet together and pushing with all her might, but we were winning the battle. When she would strain to her maximum, her legs would start to buckle.

"Jab her in the ribs and pound on her back," I half screamed at the boy, as I eased up on the pressure a little. She re-adjusted her feet, stopped the push and gasped for another breath. The ends of the uterine horns were all that remained and I knew with a good steady push I could get it all inside, if I could hurry and get it done before she regrouped for another push. I took things in both hands with fingers along the sides so I wouldn't poke a hole in it, and pushed with all my might.

It was starting to slide in when she bore down again with all the power she could muster for several seconds, then flopped over on her side. Out came the whole thing in the mud and grime. If ever a person has the urge to kill, it's in that second when they flop down. All that effort for nothing, but there isn't time to cry about it.

"Grab that other rope and put it on her back feet," I barked as I put one knee and my hands on her hip to hold her down. "Now pull them back out behind her." The two of them pulled the feet back, and I took her tail and rolled her up on her brisket. "Now leave that loop on one foot, and tie the other end of the rope to the other one." I gave the orders as I untwisted the bleeding uterus. "All right, tie her legs to the fence in case she struggles." I pushed the legs apart a little and knelt down between them.

"Get me those other two jugs of water out of my truck and put that pan affair under her uterus as I hold it up, and we'll start over," I said resignedly as I picked up the dirty uterus. "Now if you two will get on either side of her to keep her straight up I can put it back in. It's a lot easier when they're in this position. We were so close if she hadn't gone down." I shook my head. I began to wash and turn and wash some more. When I felt it was ready, I rinsed my arms and started the process over. She strained some as I was working it in, but in that position she wasn't able to generate nearly as much pressure, and I was pushing partly downhill. It made a lot of difference, and I soon had it all inside. I had to reach in up to my shoulders to straighten out the uterine horns. Then I sutured her, so she couldn't get things out again.

"Okay pull the ropes off and get her feet back under her." I started checking jugs for some water and picking things up. There was only a little water in one jug. I put it between my legs and tipped it to wash my hands. The blood on my arms was about dry.

"I can get you some water from the river," the boy offered.

"No thanks. This will be good enough. I got my hands clean. I can drive and I'll wash the rest off when I get to the clinic." I was sure wishing it would stop freezing so I could carry water in my little tank in the truck. I had bloody coveralls and my arms were covered to my shoulders, but it was getting dry fast in the nippy breeze. I put everything in the truck and headed for the clinic and my next cow. I wondered how she was doing. I'd been a little longer than I planned.

I was cruising down the highway east of Duchesne feeling the urgency to cover the fifty miles to get to Tridell, when I looked in my rearview mirror to see a red light on me from the car behind. I glanced down at my speedometer as I pulled my foot off the gas. It was reading a mile or so over seventy. There was no question he had me. I was speeding, and that meant a ticket. What else could go wrong today? I spotted a wide spot at the side of the road and came to a stop and rolled down my window. I sat and waited. Keith Hooper got out of his patrol car with his book and walked up to my window. He took a look inside.

"What in the world happened to you?" he sort of gasped as the look on his face suggested I was ready to topple over on the steering wheel. I hadn't thought about how I looked until I saw his face. The situation had an amusing aspect and a smile flicked across my face momentarily.

"I've just been working on a cow and I didn't have enough water to wash up." I could see the calm come back to his eyes as he recognized the situation.

"What did you have to do?" I guess he was wondering how I got that bloody, but then everybody is curious about vet work.

"She was a prolapse, and her uterus was all turned wrong-side-out after calving. It was one of the Rasmussen cows up the Strawberry a little ways. We had a hard time getting it back in. Took a little longer than I expected, and I ran out of water. Anyway, I've got another one with the same thing in Tridell as soon as I can get there. I guess that's why I was pushing it a little. I hoped I was urging him to hurry and get on with it. I needed to go.

"Well I won't hold you up. Drive carefully, and for hell

sakes try to stay within ten miles of the limit." A broad understanding smile lit his face.

"Thanks. Thanks a lot. I'll watch it," I said as I started the engine and pulled it into gear. It was cold and windy. I was bloody and stinking and I had another one to do, but it was a great day. I felt on top of the world.

12
SOME THINGS I LEARNED

The Pecking Order

Sometimes knowledge comes to you in slow motion, like a sunrise on a clear day. You have basic knowledge that has a relation, but it takes some time and assimilation to bring things to light. This was the case when I learned how important a simple thing like a pecking order could be in a cattle herd. The Ute tribe had about fifteen-hundred head of cattle in one herd up in Coyote Basin for winter feeding. The snow was well over a foot deep covering almost everything the cows might find to browse on. They had involved some nutritional expertise to formulate a ration and were weighing the feed to assure that an adequate amount was being fed. They were feeding on trampled-down snow, and there was very little waste and it was being compensated for.

They came to me with concern about possible parasitism, because some of the cows were getting thin. I had them bring me some stool samples to check for parasite eggs, and they were all negative, so I told them I didn't think it was worms, but there must be some reason. I offered to go up and do a necropsy if any of the cows died, to see if we could learn from that what the problem might be. It was in February when they called me to come up and do a necropsy on a cow that had died during the night. When I got up to the feeding area in Coyote Basin, it was just after noon, and I drove through the herd basking in the sun. I could see some thin cows, but didn't see anything that rang any bells as to why. Most of the cows were in good shape. Just as I returned to the bottom end of the mile-long herd, the feed trucks arrived and began to spread the hay on the packed-snow feeding area.

I decided to do the necropsy while they were feeding, and when I opened the cow up, there wasn't one bit of fat anyplace. The liver was normal without evidence of fluke, and I couldn't find any worms at all in the stomachs or intestines. I thought the cow must not be able to eat to be this emaciated so I followed the esophagus up through the chest cavity and checked the teeth. Everything was normal, so far as I could tell, except the emaciation. When I opened the pericardium around the heart, what was once fat on the heart was a reddish gelatinous mass. It was typical of what I'd been taught was sufficient cause for condemnation of carcasses in my meat inspection work. The cow had literally starved to death and I explained it to the man who was there helping me. He assured me that it just couldn't be, because they were weighing the feed every day. Something was wrong, and I wanted to know the answer.

I began to walk up past the cows as they were rapidly consuming all the feed they could. As I walked along, I noticed that all the now-evident thin cows were segregated out to the perimeter. When one of them would start to come in after some feed, one of the fatter cows would bunt her away.

The light began to dawn: the bossier cows were keeping the more timid types away from the feed, and as they got thinner and weaker, the problem increased. I got all the men together, then went along the herd showing them what was happening. I suggested that they cut out all the thin cows and feed them over across the fence by themselves. There were no more deaths, and the thin cows began to put on weight.

It took a big herd and two hundred thin cows to point out the problem for me, but the lesson remained and I saw it repeated many times in all sized herds. Sometimes it would only be one cow at the end of the pecking order, but if the owner would feed her separately, it made a world of difference.

Never Give Up

Many of the things I learned came as the result of my not wanting to admit defeat, and not having been told it was impossible. People's confidence and willingness to try helped out also. I remember the first dislocated hip in a cow. Dee Allred and I went up to treat one of his cows that didn't feel well and hadn't cleaned. He took his horse and I followed him over into the boondocks on the Uintah River. He sent me out through a pasture and said he would go around where the cows were and bring her up to the fence. The cow wasn't far from the fence, and he soon came to rope her. She was a bit thin and gaunted up, but showed a lot of spunk when he approached to rope her. She started to run, and when he caught her and stopped the horse, it flipped her off her feet right on the ditch bank in front of me. I hurried through the fence and grabbed her front leg to hold her down while Dee stepped off his horse and made a halter of the rope so it wouldn't choke her. He tied the rope to a fence post, then took my place holding the leg back while I got some water and pills and removed the placenta. After I had given her a shot and we removed the rope, she had a hard time getting to her feet. As she limped painfully forward a few steps, I could tell that we had injured one of the back legs. Dee hurried and dropped the rope on her again and she was plenty willing to stand still and let me examine her. I could soon tell that her hip was out of joint, and what to do was a real dilemma. There she stood, a little two-year-old heifer, rather thin, somewhat sick and her calf trying to fill his tummy from a rather empty udder. I realized she couldn't be salvaged by slaughter and with a leg as bad as that she wasn't going to fare very well in summer pasture. I'd never heard of trying to put a hip back in on a cow, but I had done it on some dogs. My mind was exploring as I measured with my hands from her pin bones to find the head of the femur over the top and forward of the socket. To get a dog's leg back in, it took all the strength I had. How could we do it on a cow? Maybe the horse and rope. I didn't have much hope in that. It wouldn't be steady and I wouldn't be able to manipulate the leg. My calf puller. The thought flashed through my mind at about the same time Dee asked what we ought to do. I explained that I had never tried to put a cow's hip in, but wondered if it could be done with a calf puller. He was willing to try.

I brought the puller and a long obstetrical chain from my truck and explained that I couldn't find anything for padding except my coat. He walked over to his horse, pulled off the saddle and brought the saddle blanket back. We wound a rope around her front legs and pulled them together. She laid over without much struggle with the injured leg up. I slid the two looped ends of the chain over her foot and up past the hock, then put the padded britchen of the puller between her leg and her udder, against the bottom of the pelvis. This made it possible to pull on the leg with the jack and exert a lot of pressure. The cow showed some pain, but far less than I anticipated. When I could feel that the bone had moved toward the socket, I tried manipulating it by twisting the leg. The cow showed considerable pain, so I swung the puller forward a little and had Dee hold it. I then did a quick hard twist by pulling up on the hock with one hand, and pushing down on the stifle joint with the other. I heard a muffled snap as the head of the femur popped into the acetabulum. It looked right and when we released the jack and moved the leg, it seemed to function right without any pain. We took off the chain and the ropes and helped the heifer to her feet. She walked off cautiously, like it was sore at first, but the leg looked normal.

I was very elated at the outcome and have tried many more times with other cows. Only about half have been successful and I could never understand why they all didn't work. I know when I could get one back in, it was a thrill to me and much appreciated by the owner. Another case I remember well was a big holstein dairy cow of Elvin Thacker's at Upalco. When she walked off, he literally jumped up and down for joy.

Yummy Tails

"Basin Veterinary Clinic, Dr. Dan speaking."

"Dr. Dennis, are you still coming over to the Vernal area one day a week?"

Yes. I come on Friday. What do you have for troubles?" I didn't recognize the voice, but I didn't know a lot of the people from over there.

"I've got two palomino saddle horses and the hair's all falling out of their tails." The first thought that crossed my mind was parasites: lice or mange.

"Are they rubbing and scratching a lot?"

"No sir, that's the puzzling part of it. I've never seen them rub at all, and I can't see anything on them either. By the way this is Milt Sagers." I wondered if it could possibly be photosensitization. I'd better have a look.

"Will you be home Friday?"

"I probably won't. I've got to be at the store in town, but the wife will be here. She can show you, one of them is hers anyway." Some women were good at handling horses, and then there were others I'd just as leave not have around.

"Do you come home for lunch?" If I could catch him there I'd prefer it.

"Yes I do, it's only a half mile or so."

"How about if I come during the noon hour?" I usually had

the more urgent things under control by then.

"That would be great. Do you know where our place is?" I didn't have a clue, but I was good at finding places if they gave me a direction or two.

"No. I'm afraid not, you better give me some directions. Paint me a picture in my mind, and I'm sure I can find it." He told me how to get to his place. It was out west toward Maeser.

On Friday, about a quarter to one I was approaching the place and I could see the two palomino horses and a couple of goats in a ten-acre pasture that looked lush and green. Even from a distance, I could see the bare tails. I pulled into the yard and he and his wife both came out.

"I'll go get them in. I just call, and they come on the run when they see the grain bucket," he said. He entered a little shed and emerged with some lead ropes and a bucket of grain.

"Have you changed feed lately?" My mind was still blank as to a cause.

"No. It's just that pasture. We do give them a little grain when we want to catch them, but it's the same batch of oats we've had for months." He looked at me inquiringly, wondering, I guess, what was on my mind. I was thinking I might have to walk over the pasture to see if I could find something. I'd better have a look at the horses first.

"Well, let's have a look. I'm still puzzled over what it might be." He began calling and shaking the bucket as he walked toward the gate. The horses wrung their little bare stubs as they bucked and ran toward the corral. He came back over by the fence where I was standing inside the corral. One horse came right in and walked up to partake of the oats, while the other lingered outside the gate. He must have been a little leery of a stranger. They were a beautiful, well-matched pair that looked the picture of health, except for the tails. He clipped the lead rope on the halter and handed the bucket to his wife.

"My wife and I love to ride, and we go for a little while each night after work. The evenings are pleasant and we really enjoy it, or we did until this happened. Now we're so embarrassed, we don't want anybody to see them." His wife took the remaining oats over to the other horse and petted his neck as he finished them off. The two big goats made their way in after the horses and were standing around looking curious, as goats usually do. I petted his horse and cautiously made my way back to his tail. He didn't care what I did, or even move when I grasped the tail and rubbed it. I looked closely for lice and mange, or abrasions from scratching. There wasn't anything. The skin looked as healthy as could be, only it didn't have any hair on it. One thing that caught my eye was several inches of normal hair at the top. It was cut off as straight as a girl's bangs. Both owners were standing close, watching me now, although they were silent, I could tell they were anxious for my diagnosis, which I didn't have. I decided to have a look at the mane for some clue and moved up there. When I didn't find anything I wondered if the other horse might have some evidence I couldn't find. I looked out to it, and there was one of those goats reaching as high as possible and nibbling at the ends of the hair at the top of the tail. Realization flooded my mind, and I stepped back, I guess with a smile. It was hard to not laugh right out loud. They were so intent, and all ears to hear what I had found.

"Do you know what it is, Doctor?" It was the wife who asked the question.

"Yes, I do now. I think if you take those goats out of your field your horses will have beautiful tails again." I pointed to the other horse as extreme puzzlement showed on their faces. The goat was reaching as high as he could, and almost looked as if he was going to stand on his hind legs.

"What about the distemper?" The question was already formulated before he saw the goat and made the connection, and I realized why the goats were there now. There was a popular belief among horsemen that keeping a goat in with the horses would protect them from distemper. To me it was like an old wives' tale and had no validity from a medical point of view.

"I don't think goats have anything to do with distemper in horses. It's a bunch of malarkey as far as I'm concerned. I think I'd rather have pretty tails on my horses."

Perceptive Dog

Steven Utley was a lad that I became well acquainted with when he was just a small boy. He grew up in the community and he started calling me about the time he started school. He would call me about birds, cats, dogs and everything else. I learned that his mom was having a difficult time trying to make ends meet and keep a roof over their heads. He would bring injured birds or dogs or whatever to my home and I would do what I could for them, but I discouraged him as much as I could from obligating himself and his mother over animals. I knew he must be a true animal lover, because he just kept calling or coming in with this problem or that, or wanting shots. He wanted me to let him work at the clinic to pay his bill, which was only a fraction of what it should have been. I appreciated his willingness and his acknowledgment of his obligation, but I felt he was more motivated by wanting to be around the animals.

Most of the time I didn't mind. He would use a lot of my time, but there were times when I was busy and he could be a real pain. He always wanted to know if we had dogs in the pound, and if he could have one. Knowing his mother's situation, I wasn't too cooperative. Whether it was right or wrong, I tried to discourage him, but he did acquire some pets over the years. When he was starting into his teen years his mother finally consented to let him get a black Labrador retriever puppy, and he was anxious to give it the best of care. He would bring him down by my home for me to see and keep track of. He grew up to be a fine-looking, big old happy dog, and he and his owner were good for each other.

Some years went by. Then one day Steven showed up at the clinic with his dog that had been hit by a car and had a broken leg. Our usual procedure was to put a steel pin into the bone to hold it while it healed. When we examined him, we found the bones were fairly well aligned, so we opted to use a Thomas splint which was much cheaper. The dog was anesthetized, the bones set and the splint put on. When Steven came the next day to pick up his dog, we advised him to bring him back in four or five days so we could check on the splint and see how things were doing.

When he came back the next week, the old dog didn't hesitate at all. He came right through the door, wagging his tail, walking on his splint, and feeling right at home. We checked his toes for circulation, his upper leg for pressure sores, made a few adjustments and added tape to the end that dragged on the ground. We told Steven to watch for sores at the top of the splint, keep the end taped well because he dragged it, and he would heal in about six weeks. We told him we thought he could watch it at home and to call us if he had any questions. Then he left.

About the middle of the next week we looked out, and there was the old dog waiting at the door. Steven wasn't around and there wasn't any car or person that we could see. The old dog had come on his own, hobbling along with the splint. We let him in and checked him over, taping a place or two, then opened the door, and soon he was gone. From then on, he would show up on the front lawn every four or five days and wait to be let in for his check-up. When the time was up, we took off the splint and he never came back again.

Grain Founder In Horses

After about six years of practice here in Roosevelt, my good friend, Kent Nelson, called one afternoon and informed me that his horse had just eaten about a half sack of wheat. He had come home and found where the horse had reached over the fence and lifted the chicken feed over to his side and proceeded to eat it all. I knew it was a very grave situation and rushed right up to treat him.

Grain founder was a situation with which I felt very uncomfortable. We hadn't been given very much information on it in school, and in my many inquiries to other veterinarians,

I hadn't learned much more. There were remedies used by some, but none of them were a sure-fire, corrective measure. Mineral oil and antihistamines were the standard most everyone used.

I gave the horse two quarts of mineral oil and some antihistamine, and left some to be given later. We did all that we could but the horse died anyway, and it really bothered me. I'd already had a few cases of grain founder in horses, where they had suffered with laminitis and then I was called, but this was the first one I'd known about from the first hours. Nothing seemed to really work. I thought about it a lot and I asked at every vet meeting or seminar I attended, and although there were things that helped there was no sure-fire cure. I remember my father telling about coming back from a trip to Price in snow three and four feet deep. They didn't have hay enough for the horses so they fed them just grain alone. I also remembered that when we were working the horses heavy on the farm we gave them a lot of grain, and it didn't seem to hurt them. My gray cells began to work and I thought if I get another call on a horse that has been in the grain I'm going to try another shot in the dark and have them ride it hard for four or five hours. Luckily the next one I got was a call from Newell Christensen, who worked on the forest and was used to riding a lot. I told him to get on that horse and just ride the hell out of it. I told him to give it all the water it wanted, but to really work it hard like climbing hills. Then the next day if it was gimpy on its feet to stand it in cold water. The horse never showed any founder symptoms at all, and the next day it only acted a little stiff and sore in the muscles and somewhat tired. From that time on, if I got a call on a horse that had eaten too much grain, I recommended a half a day of as hard a work as the owner could give it. It's been the best treatment I've ever heard of.

Animals Can Be Dangerous

Veterinary medicine does have some dangers associated

with it, like mean bulls, mad cows, wild horses and biting dogs and cats, but sometimes there are dangers that you don't expect.

When the local county fairs and 4-H shows wrote up their health requirements, they almost always required a blood test for brucellosis. This was wise for the breeding stock, but for the fat steers that were to be sold and slaughtered, it was a foolish requirement. At any rate, it was required, and I traveled many miles and bled them, without any charge most of the time. The bad thing about it was that the boy or girl was seldom home when we would go out to do them during the day. Mom was there to show us the critter, and that was about all. There was seldom a chute or place to work them, so the end of a long rope was the normal pattern. When we strangers entered the corral or pasture, they forgot all that gentling down, and just about reverted back to the wild. We'd have to rope them, snub them up to a post, then put the nose tongs in their nose and try to get a tube full of blood.

My father and I started early in the forenoon to go from place to place and ended up over in the Vernal area in the afternoon. We stopped at a farm and the mother showed us this big, fat steer. I think he weighed about eleven hundred pounds. I roped him and got a dally on a post in a pole fence, and Dad took the rope on the outside to take up the slack, while I moved up on him. When we had him up within five or six feet of the fence, I took the nose tongs and made a one-handed grab at his nose. That steer let out a beller and jumped forward pinning my back to the fence with my head bent back over the top pole. My head was between his front legs, which were over the top of the fence. Dad saw the predicament I was in, and knew that my neck would be broken if his weight came down on me. It frightened him, and he put his hands on each front foot and pushed that steer backwards, with a little help from me. I was able to get my head out and crawl out from under him. Just a tiny bit more pressure, and it could have broken my neck and maybe killed me. It scared us so bad, we just had to quit. I didn't get any blood, and I vowed I wasn't going to bleed any more of them, unless there was a chute.

Another time I was pregnancy testing some big, tall Brahma cross cows in a corral over to Myton for Clive Sprouse. There was a catch pen behind a rather long chute made of 2x6's. I was all alone, since no one showed up to help. There were only about thirty head to do, so I decided I could work them by myself. I blocked off the end of the chute and put about half of them in the catch pen, then I filled up the chute and started palpating on the back end. I was about halfway down the chute with about five head done and backed out into the catch pen. I just finished palpating a cow, marked her and stepped back to climb the fence and get her to back up. I heard something behind me, and one of the cows I'd already done was coming down the chute like a freight train. I expected her to hit me with her head and throw me up in the air so I raised my arms toward the top of the chute. She just blew a little snot at me and pushed right on by, squeezing me against the side of the chute and spinning me like a roller bearing between her and the side of the chute. The pressure was so great, she forced all the air out of me, and I've wondered if it didn't cause my hiatus hernia. I was glad she had room or made enough room to get all the way by me, because if she had progressed part way and then stood there pushing, I couldn't have breathed. She would have smashed my pelvis and probably broken some ribs. It really hurt, but it was over so quick, I didn't suffer very long. I think I turned two complete turns as she rolled me.

Hemoglobinuria from Thirst

Learning something that had never been mentioned in school seemed to be the name of the game for me here in the Basin. I remember one late afternoon that I got a call from Charles Crozier. He had some calves with bloody urine, and two of them were dead. It was a hot afternoon and I rushed up to Neola to check things out. As I listened and asked a few questions, I pieced together a history. The first thing I asked, since they were in his catch pen where the cows went into the barn, was where had they been before he brought them into that pen. I learned that they were kept in a small pen under the shed during milking time, but it was crowded so they let them out into the catch pen between milkings. I also learned that someone had forgotten to put water in there for them that morning. In the afternoon when his wife usually gave them more water, she found them dry. She got the buckets and water hose and gave them water, and they had been thirsty, she'd said. By this time, I'd observed one or more passing a deeply pink-tinged urine. When I checked the temperatures, all were in the normal range. I decided I'd better open up the dead ones and see what I could learn. I knew it wasn't infectious red water like I'd seen before, but they did have red urine. When I opened up the dead ones, each of them had red urine in the bladder, but everything else in the abdominal cavity was normal, so far as I could tell. I did find, when I opened the chest cavity, that the lungs had large areas where it looked like vessels had ruptured and bled out into the tissue of the lungs. In fact a bloody froth in the air passages had been the cause of death. I didn't know the answer, and I told Charlie it was a puzzle to me. It was time to milk the cows, so we dragged out the dead ones and put the others back in their pen. They went to eating and seemed healthy except for the urine, so I told him to call me in the morning. Next morning he called and said they were eating, the urine had cleared up and everything seemed normal. I don't know that I ever did get a scientific answer, but over the years, I heard about and saw the same thing happen when young cattle got extra dry and then drank all the water they wanted. I judged that the blood vascular system was stressed with the sudden intake of water to the point where capillaries broke down and hemorrhaged out into some of the body tissues, and that blood was then carried to the kidneys for disposal.

Preg-Testing Mares

I think the most significant aspect of doing the rectal

exams on colic cases was the learning experience it was for me to recognize and identify the uterus and ovaries of the mares. Since pregnancy diagnosis in cattle by rectal palpation was becoming an accepted and worthwhile diagnostic tool, I began to wonder why we couldn't do it in mares. Consequently when I was asked if I could tell if a mare was pregnant like I could in a cow, I said I thought I could. I then began to palpate a few mares, and could indeed feel colts in them and uteruses that had fluid in them. Since early detection was what the owners wanted, I started doing them at about two months. This was even easier than older pregnancies, and detecting an open uterus with nothing in it was often easiest of all. Within a few years they were teaching the technique and other practitioners were doing it. It began to be talked about in seminars, taught at some of the vet meetings and much knowledge was shared, but for me, it was a shot in the dark I took on my own.

Word got around that I could tell if a mare was pregnant at about two months and this was about as early as the blood and urine test could be done, so I was called on to do quite a few. I always liked to have some poles or a low fence between me and those hind feet, and even then, they kicked over the top sometimes. Once in awhile, I would do a real gentle mare out in the open by standing to the side when I first entered the rectum, but this wasn't always wise. One day while I was vaccinating calves at a man's place, there were three or four kids riding and climbing around on an old mare that didn't seem to care a bit. When we got through with the calves, the owner asked me if I could check the mare to see if she was with foal. I got my glove on and lubricated my hand and arm and approached the old mare while he held her. I stood to the side some and petted her as I spoke to her and took hold of her tail. She didn't make any fuss as I pushed my hand into her rectum and began pulling out the fecal material. Then as I inserted my arm farther and began to feel for the ovaries, she came up with both hind feet and sent me sprawling back on the ground. Her feet hit me just below the belt on the hip and in the groin as I turned sideways. Luckily I got turned as much as I did and falling away helped to keep me from being hurt too bad. I learned that you can't trust even the gentlest mare.

On another occasion I went up to Neola to test some mares for Dick Bastian at his dad's place. They had a good narrow alley to put them in and we could put poles across to protect me. I think there were six mares and I could stand on a catwalk at the side of the alley to start, then slide over the fence and finish down behind the poles. Things went well for the first five, but the last one was a mare that was only halter broke. She had never been worked with or ridden, but she raised good colts, so Dick kept her for that. He warned me she was a snake and that we might have trouble. She walked right into the alley and they put the two poles behind her, then backed her up to them. When her legs hit the poles, she lunged forward and lashed out with both feet, but eventually backed up and stood there. Now it was up to me. I petted her and got hold of the tail over the fence and raised it up. When I touched the anal opening, that old gal kicked with both back feet so quick her hocks hit my hand before I could get it out of the way. I tried several times and she could hit my arm every time, but it was so high up on her legs that it didn't hurt me. When Dick said she was a snake he described her well, because those feet were about as fast as a snake striking. We gave up that day, but a few days later he had moved his horses down to a place out in the cove, and I went out there. We just flipped a needle in her neck and gave her a general anesthetic. When she went down, I laid down on the ground behind her and checked her that way. I wasn't sure I could feel things the same as when they were standing, but I was able to tell that she was pregnant. She was the only one I ever had to put to sleep to test. I thought if it was important enough to the owners, I was willing to go that route.

De-Scenting Skunks

At about the time I graduated from veterinary school there was a new fad sweeping the country, at least in the intermountain area. It was the removal of the scent glands from skunks at an early age, and then making household pets out of them. Over near Denver, there was even a skunk farm where they were bred and raised for this purpose. They did seem to respond and make fairly good pets, and with their contrasting white and black colors they were rather pretty as a sleek shining animal in the home. Their popularity was really more of a novelty that certain individuals got a lot of enjoyment from. They frequently caused temporary shock when strangers encountered them, especially when startled a little, they retained their natural instincts of going through the motions of spraying. In the basin, people would find the baby skunks and think they were so cute they wanted to keep them.

To remove the scent glands was a bit ticklish, but not a difficult operation, and in my earlier years of practice I did quite a few. They were brought to me as babies without well developed scent glands and a minimum of odor. It was not wise however to bring them into my clinic in the home, because the tiniest amount of that liquid was an extremely powerful "perfume." I would do them outside on the endgate of my truck, which I covered with several layers of newspaper. I would wear two pairs of surgical rubber gloves, and a smock and coveralls that I could remove. I used an injectable barbiturate drug in the abdominal cavity for anesthetic, or more often just ether, because it was safer. The barbiturate anesthetic gave me more time, but it was difficult to judge the dosage. I usually saturated a pledget of cotton with ether and dropped it in a small tight box with the skunk in it, and when they were anesthetized, I'd work fast. Sometimes they would start to come awake and begin to move before I was ready. I usually put their heads back in the box with the cotton and ether and had plenty of time.

On either side of the anal opening, half in and half out, were two small nipple-like nozzles. I would grasp that nipple with some forceps and carefully dissect down around the outside of a sack enclosed in muscles. It was imperative to keep the sack intact and get it all out, allowing the muscles to contract and stop the hemorrhage. They were healed in a day or two, and all was well. There was enough smell on their fur from being around their mothers that handling them would contaminate the rubber gloves until they couldn't be washed off successfully. I would peel them off, and wrap them and the removed glands in the newspaper, and burn it all in a small bonfire. The fire removed all traces of the odor.

Just to give you an idea how strong that scent is, I spilled one small drop on the newspaper while I was operating in my open garage one evening. The next day I heard some families talking about smelling a skunk about six the evening before. They lived downwind from my place four blocks, so you see how far one drop can go.

After about ten years, the National Veterinary Medical Association began to caution us about the dangers of having skunks and other exotic animals for pets because of the danger of human rabies exposure. I didn't take the warning too seriously, because we had never had rabies in this area, but I had my eyes opened by the following experience.

I operated on a baby skunk for Mr. and Mrs. Grant Hansen and they house trained it and kept it in their home. It was quite a novelty and a lovable pet that roamed their home at will. One day when the mating season and hibernation time came, the skunk disappeared. It was gone for five or six months. Then one day when Rita had the kitchen door open and she was out in the yard, she turned to see a skunk going in the door. The anxiety really began to build when she could smell the odor. Was it the pet she had lost, or was it another hungry skunk? She didn't dare try to scare it out of the house for fear it would spray, so she just kept watching it. Finally from some of the things it did, she concluded that it was her pet skunk, although it had some odor on its fur from being around the wild ones. When they told me about the experience, I suddenly realized how easy it would have been for that pet skunk to walk into that house carrying the rabies virus. If we had rabies endemic in our wild animal population this could be a serious danger. Shortly after this, there was an outbreak of rabies in Sanpete

County, and their greatest problem of containment and eradication was in the skunk population. From that time on, I never operated on any more skunks.

Ear Problems

So-called ear warts in horses made them so sensitive about being touched around the ears and face that putting on bridles and even halters would be almost impossible. They would throw their heads and rear up and even strike with their front feet if the owner tried to put a bridle over their ears. It was a real problem and it ruined some good horses. Once I had my chute where heads could be restrained, poles put behind them to prevent backing up, and we were higher than the horse on the catwalk. I began to examine some of those horses. I concluded that it wasn't anything like true warts, but seemed to be more like a fungus growth, or a burrowing mite. I did some skin scrapings by tying the head and using a twitch on the nose, but I didn't find any mites. To me it had to be a fungus infection, so I began treating them as such. I tried several medicines and different approaches, but scraping off as much of the dandruff-like scaly covering, then rubbing them vigorously with the original old stinking KRS solution worked the best. It was sometimes a real fight to treat a horse. Some would throw themselves in the chute, and all resisted about as much as they could. The KRS was a fly repellent with bone oil and several other ingredients, but it worked well for this condition. I think it worked well because it had such a capacity to penetrate the tissue. You couldn't wash it off your hands, or the smell from your clothes. You just had to wear it off, but it worked well, and the owners could follow up by squirting it in the ears. In a few weeks, the horse would let you rub its ears again, and all was well.

Emphysema of a Pet Dog

There was a little dog brought into the clinic with a history of a chronic cough. I checked its temperature, palpated the larynx and trachea, and looked down its throat. The throat looked a little irritated from coughing, but otherwise things were normal. When I listened to its lungs with my stethoscope, I could hear some rales and a wheezing noise. The upper part of the lungs were entirely silent. I told the owner that the dog had emphysema and discussed some possible treatments. They said to go ahead and do whatever I could. When I gave him some epinephrine, there was obvious relief and my diagnosis was confirmed. I asked them about anything he might have come in contact with, or a change in his lifestyle or exposure to dust or aerosols. Nothing had changed, he just breathed harder and coughed a lot. I gave them some pills to give each day and sent him home for a week's treatment. When I questioned them, no bells rang in my head, but after he was gone, I kept thinking about the poor little fellow.

In a few days, I called them up for a progress report. They said they went visiting the rest of the day after I treated him, and he was just fine, just coughed a little if he barked. The next morning he was coughing again, so they gave him his pill. They felt it helped some, but each day he had gotten worse and the pills didn't seem to be working. They wondered if they could bring him in and leave him for awhile and have me treat him. Maybe I could clear it up. From what I'd heard in his lungs, I didn't have a lot of hope, but I felt obligated to try. In three or four days, he was surprisingly improved. They were going to be out of the area for about a week, so they stopped to see him and left him while they were gone. We didn't give him any more treatment, except tender loving care, and by the time they returned he was frisky and didn't cough at all.

The recovery was more spontaneous than the result of medication, I felt, and I couldn't get the thought out of my head that it was something he was encountering at home. When they came to get him, I noticed both the husband and the wife took deep drags on their cigarettes before stamping them out prior to coming into the clinic. When they came in the door they just reeked with the smell of tobacco smoke, and the cogs began to turn in my head. I had a suspicion of what was causing the dog's emphysema, but I now had a problem. Would I offend them if I suggested that it might be the tobacco smoke? I explained how he had recovered there at the clinic, and told them I thought it must be something he was exposed to at home that triggered his problem and casually suggested it might be the smoke. I told them if he started to cough, to call me. It was only the next afternoon when they called to say he was wheezing and coughing again. They also volunteered their opinion that it was the smoke causing the trouble and confessed to both being chain smokers. I suggested they build him a dog house and put him outside, and see if he didn't improve.

It was about two weeks later when they brought him in again, and he was having more difficulty than ever. They explained that he had always been a house dog and they couldn't bring themselves to kicking him outside. They said it would hurt his feelings too much. "We've decided to put him to sleep."

Didn't Bother to Look

"Hey, Doc, when is pen 105 gonna sell?" It was Reed Mckee and I was sorting and penning the cattle to go through the auction ring. I stepped out through the little gate and into the little covered room with a stove that served as our office.

"It's the next pen up, right here on the top," I indicated as I held up the sale tickets. Reed had followed me in.

"Good, my cow's in there and I need to get home to tend my water." Reed worked as hard as anyone I knew, but he always had a little more to do than he could get around to. I think he had thirteen kids and every one of them were workers, too, from very young ages. It seemed like he had just a few too many acres to farm, a few too many cows for the barn, too many calves in the pens. Routine vaccinations, dehorning, and castrations were a week or two late. He was really a good man, but he was always behind the eight ball, as they say.

"They're bringing pen 105, it won't be long now," I assured him.

"By the way, Doc, I got a lame calf up there that ain't gettin' any better. I usually have a little foot rot in my calves. They limp around for a week or so and then get better. I don't usually bother to treat 'em, but this calf has had it for awhile and I think it's getting worse. What should I give him, some penicillin?"

"Well it might help, but I think you'd get better results with sul-met. At least I've had better results with it. I have used sulfa-pyridine, it does a better job, but it's more expensive."

"Have you got some sul-met with you?" I always kept my truck well-stocked when I went to the sale on Saturday. The farmers would buy things from me at the sale and never stop at the clinic.

"I've got the liquid, or the pills." He was following me out to my truck.

"Give me the pills, I get along better with them." I pulled the side door down on my truck and laid out a paper towel. It was something I could wrap the pills in, and also something I could lay out the dosage for each day on. I figured it helped the clients to remember better.

"What does the calf weigh?" All medicine like this was dosed by the body weight of the animal.

"Well he was about three hundred pounds, but he's fell away some since he got lame." I quickly pictured the animal in my mind, probably have trouble swallowing the big pills. I got the bottle of medium-sized ones.

"Here, give him five of these the first day, then three the second day and the third, then two on the fourth." I laid them out a day at a time on the towel.

"Maybe I should take the whole bottle?" he queried. That's how Reed was, he always wanted a little more than he really needed. He also had a tendency, like a lot of other farmers, to think that if a little is good, then a lot is better. With the sulfa drugs, this was a dangerous thing, because if the animal was overdosed or treated too long, it would crystallize out in the kidney and cause death. I was a little worried and the dosage on the bottle was a little hard to understand. I needed to impress upon his mind the importance of treating the animals properly. Yet, if he had something, he might treat some of those others that got lame.

I finally said, “It’s up to you.” Maybe I didn’t show much enthusiasm for selling the whole bottle, or he could read the concerns I had by my eyes.

Somehow he sensed my hesitation and said, “Oh, I’ll take these and try it on him, and if it works, I’ll buy a bottle then. You say five the first day, then three and three and two?” I nodded my head in agreement. “I’ll tell the kids.” I already knew the kids pretty much cared for the calves. He asked to charge the pills, and I went back to work.

My responsibility at the sale was to observe all the animals to see if any were sick, as well as vaccinating the pigs and vaccinating and testing calves and cows that went back to any farm. I could walk down the rows of pens and observe, but it wasn’t easy to pick out a sick one in a whole pen full. Another option was to sit in the sale and watch them go through the ring. Here they were kept moving, and turning, but once in there, it was too late to alert the owner or the sale barn operator. I felt the best place to check the animals was while they were being separated into pens before entering the auction ring. This, then, became my routine and since it was hard for me to just stand and watch, I soon got involved in helping to sort. For many years, then, I worked in the cutting pen. I wasn’t on the payroll, but I had to be there, and it was easier for me to work than to sit around. All of the farmers and stockmen soon knew where I was when they needed my advice, help, and medicine on Saturday.

On another Saturday, it was Richard Olsen who came out to the cutting pen to get me. “Dr. Dennis, could you come and look at a cow I’ve got in my truck. She’s got a lump on her jaw. She’s the one I got that sodium iodide for at your clinic a week or two ago. I gave it to her in the vein like you told me to, but it hasn’t helped her any that I can see.” Richard was one of my closest and most trusted friends. When I began the brucellosis testing program it was Richard that I’d hired to help me. I’d taught him how to bleed the cows, vaccinate and tattoo calves, fill out the forms and keep the records. He'd helped me for many years before he was hired as a federal technician. I thought I might be able to lance the lump and drain it, so I stopped at my truck and picked up my nose tongs.

"Has she got a calf?" I asked Richard as we approached the truck. I was thinking of giving her some more sodium iodide as well as lancing the lump, which wasn't recommended when they were heavy with calf.

"Oh yes, she had her calf before I gave her that medicine." He took the tongs and stepped up on the back of the truck where the cow's head was. I think he was the most willing worker that I had ever known, unless it was my dad. At any rate, he caught the cow's nose and dallied her up to the corner of the rack. I climbed up on the side of the truck and felt the lump.

"It isn't typical, Dick. I thought we might be able to lance and drain it, but I don't think there's anything in it. It's hard all the way through. I want to see the inside of it from her mouth. I'm going to slip down in there and have a look." I started working my way around the back of the truck to get to the other side. Dick stepped down, still holding the rope. I got over to the other side and climbed down into the truck. When I opened the cow's mouth, I got a glimpse of something shiny on the side where the lump was. I reached in and pulled her tongue out the side of her mouth and wrestled with her until she stopped fighting. When I looked again, I could see this jar lid caught outside of her teeth and cutting into the soft tissue. It was causing all the swelling. I happened to have a pair of hemostatic forceps in my pocket from pulling vessels after dehorning some calves. I reached in with them and pulled out a Kerr fruit jar lid center and handed it to Richard. "I think she'll be able to eat better now and get better." I chuckled.

"Dr. Dennis, you make me feel bad. I think that has been in her mouth for a month and I didn't think to look. It sure is amazing what cows will try to eat."

About three weeks after Reed Mckee got the pills, I was in

the office part of the sale eating a hamburger, when he came in the west door. They had pop machines in there and in one corner a little stand where a lady prepared hamburgers and coffee and sold candy bars. The office part was screened off with two countertop windows, one where you paid for the animals you bought and the other where you picked up the check for those that were sold. There was always a goodly number of people in this office area. Reed stopped to order a burger and looked over my way. I remembered the calf and the pills. "How'd that calf do?"

"Those pills never done a damn bit of good. If anything, he got worse. He's packin' that leg now." It seemed like he said it as loud as he could to make everyone in the room hear. I wondered why everybody got so much pleasure out of telling me that the medicine didn't work, that the cow didn't react like I'd said, or that they could buy vaccine cheaper some place else. I guess I was a little sensitive to their constant effort to cut me down. He came over and sat down to wait for his burger to cook.

"Maybe I ought to get some more pills and give him. He's really lame now and getting skinnier every day." I thought that if it didn't work the first time, it probably wasn't going to now.

"I'm not sure it will do much good if it didn't work before," I said a little apologetically.

"Well, I don't know if it helped him or not. I left it up to the kids at the time, but I saw him yesterday, and he's really packing it and looks like hell. I probably ought to do something. Have you got anything better?"

"No, not really. The only other thing I have done is shoot penicillin right into the foot. It's hard, there isn't much of any place for it to go, and they fight it. On some chronic cases where it's mostly on one side, I have amputated that toe." I could see by the look on his face that neither of those options interested him.

"Naw, I think I'll try the pills again. Sounds like too much trouble to me." The calf appeared expendable in his mind.

"I'll get 'em ready and lay 'em up on top of my truck bed for you." I got up and headed out the door and back to work.

It was about a month later when Reed called me up to his place to treat a milk-fever cow and clean the afterbirth out of another one. As I was washing up, I thought about the calf.

"Whatever happened with the lame calf?" curiosity prompted me to ask.

"Oh the damn thing's still out there. I should shoot it and drag it off. It's down to skin and bones. We can go take a look at it if you want." I think he was ready to make a decision about it. I took my rope and we walked to the calf pen where he had it. It was typical of his place: too many calves making the manure mucky, and this poor calf having to fight at the manger for food. I walked up to the calf and dropped the rope around its neck and pulled it back into the corner of the shed to the only dry spot. It was weak and wouldn't touch that foot to the ground. I grabbed it by the flank and front leg and threw it down with that leg up.

"Here hold him down, while I take a look at that foot," I told Reed. I moved around and picked up the foot. There was a cut through the skin all the way around it. I was pretty sure I knew what was wrong. It was either a string or a wire cinched up around there. I took out my pocket knife and opened the blade I left unsharpened for this kind of probing. I don't know whether it was my lucky day or the poor calf's, but the first probe down in that cut popped the loop end of a piece of baling wire loose. I took hold of it and pulled it out of the cut all the way around that foot. There was a tiny kink on the other end where it had been hooked. I handed it to Reed and said, "I think he might get better now."

Dagley's Two-Legged Dog

Over the years, I was amazed how dogs could adapt to the loss of limbs, or parts of them. One dog stands out in my memory over all the others. The Dagley family lived on a ranch on the south side of the Duchesne River opposite to the Randlett area. One summer they cut up their dog in a windrower and rushed him up to me. When they arrived, the dog had rags tied on three legs and the fourth was cut. He was so weak they laid him on the table and he didn't move. He had pale membranes and evidence of a lot of hemorrhage. There were three of the boys that came with the dog.

"Can you fix him up, Doc? It was getting dark and we hit him with the windrower." I wasn't as concerned with trying to fix him up as I was with determining if he was going to live. He certainly couldn't lose any more blood and make it. I carefully unwrapped each stump and clamped hemostats on any bleeding vessels. It gave me a quick survey of how bad off he really was. Two legs on the same side were cut off more than halfway up and the toes were gone on the other front leg. One toe and the skin were cut on the back foot. I didn't think he was going to be able to survive, and with that much damage to the legs, I didn't think he could ever walk if he did.

"Well, fellows, I hate to tell you this, but I think he would be better off if we put him to sleep. He'll never be able to walk if I can save him." In my mind there wasn't any other rational course to take.

"Oh no, Doc," they said. "We can't do that. It was all our fault. We want him saved. We'll take care of him even if he can't walk." I could tell they were carrying a great load of guilt, so I didn't push the euthanasia, even though I thought it was the right thing to do. My sixth sense kept telling me we didn't have a chance anyway.

"I can try, but I don't think our chances are too good that he'll survive it. He's lost a lot of blood." I didn't have much hope and I wanted them to understand how serious he was.

"You gotta keep him alive, Doc. He's too good of a dog to die." They were adamant about it. I was hoping they wouldn't blame me if he died.

"Okay, I'll do the best I can." I was always willing to do the best I knew how. I gave him some medicine for shock and some morphine for the pain and as a pre-anesthetic. I put him on a blanket and filled some bottles with warm water and wrapped them up close to his body. "That medicine will take about forty-five minutes to work, and the surgery is going to take a long time. Maybe you'd like to go home and I could call you when I get done, if it isn't too late. I don't want to wake you up if you're asleep." I thought it was best if they weren't there to watch. Cutting off bones can sometimes be a little gruesome, and if the dog died they didn't need to see that.

"We don't have a phone. We'll have to call you, but I guess we might as well go if you don't need our help. I sure hope he makes it." They each had to rub the dogs head and leave him words of comfort before they left. I was alone with my challenge and wondered how much I'd have left when I finished. At least I thought they were aware that he could die.

With the effect of the morphine and a little deadening, I was able to remove the one severed toe on the back foot and close the skin down to the side of the pad. That was going to be his best foot. Then using as light a dose of general anesthetic as possible, I cleaned the wounds, sawed off the bones and sutured the muscles and skin over the stumps of the two legs that had been cut off. I'd had to add a little anesthetic as I worked, but he was still alive. On the front leg, I had to remove some of the bone, and then I pulled the pad over the end of the stump making a padded end. After nearly two hours, he was still alive, and I think I felt a little like the boys. I wanted him to live too.

The healing was miraculous. By the time I took out the stitches after two weeks, the dog was able to sit up, eat and drink without help. It was about two months later when they had me come down to their place to treat a horse and preg test a couple of cows. When they took off on a horse to round up the cows, that dog rolled up and launched himself forward running on what he had of those two legs on the same side and kept up with the horse on a trot. When he would stop he flopped down on his chest and waited. It was truly a miracle how he could get up and balance on two legs and follow that horse.

Impossible but True

One day, George Houston brought in a calf about six

months old with a very obscure lameness. It was weak in the hind quarters and a little unstable as it walked. The symptoms were not typical of anything we had ever seen before, except it resembled white muscle disease. This is a weakness thought to be the result of impairment in the circulation to the muscles and is caused by a deficiency of selenium. The history was that a few nights before, its mother had been killed by lightning. There wasn't any evidence of burns on the calf so we treated it with some bo-se, a selenium-tocopherol medicine used for white muscle disease and sent it home. We told George that if that was the trouble, it should be back to normal in twenty-four hours, if not, to call us. The next day he called to report that the calf was no better and didn't want to eat very much. It was a puzzle, and time is sometimes the only thing that brings answers.

In the course of the next eight or nine days, the calf failed to improve, and it was evident it was losing ground and suffering. Ultimately, George decided to slaughter it and what he found was utterly amazing. The pelvic and leg bones on one side were burned black and would crumble and break with the slightest pressure. The muscles next to the bone were cooked to a chocolate color. A bolt of lightning had passed down through that back leg, literally cooking it, without killing the calf. How could such a thing happen? It seemed impossible and many hypotheses of how it happened came forth. Visualizing what might have happened, I felt the calf must have been standing with that hip in contact with its mother when the lightening struck. Enough of the electrical current passed down through the calf's leg to cause the damage, but didn't go through any other part of the body, or it would have been killed. Sometimes things are impossible, but true.

Another seemingly impossible occurrence (and a cause to rejoice) happened with a dog brought in for a spay. Our normal procedure was to give a pre-anesthetic of morphine and atropine, which reduced the general anesthetic level and dried up the secretions. It almost always had the side effect of causing the dog to vomit. Once we gave the shot, we placed the dog in an outside kennel that could be washed down. After the usual half hour waiting time, we went out to bring in the dog. There in that kennel was one of the large rubber calf nurse bottle nipples. They have a base about two inches in diameter and a nursing part about four inches long. This wasn't too big of a dog, but somehow it had swallowed that nipple at home, which I would have thought was impossible. Even more impossible, it had vomited it back up. That thing was so big it could never have passed on through the intestine and surely would have caused a blockage. Somehow it must have lined up with the esophagus and was forced out. Another occurrence one would declare impossible, and yet it was true.

Special Needles and Things

Over the years, there had been some interesting, even strange requests for needles, equipment, and medicines used by veterinarians. The word seemed to get around and people came with needs and hopes, but embarrassed and reluctant to a degree, to make the request.

There were those who wanted good-sized long needles and syringes to inject glue deep into cracks and out-of-the-way spaces in lumber and materials. Others who wanted the same to inject insecticides into holes and cracks to fight cockroaches and termites.

There were various medicines that were purchased for a lame horse, or a crippled dog, that were actually used to treat the discomfort and pain of rheumatism or arthritis. The old veterinary white linament, linagel, and rubron must have given satisfactory relief, for it was praised when refills were needed. Then in more recent years, butazolidin and DMSO didn't always go to the animals for which it was purchased.

I furnished several upholstery shops with heavy suture needles, curved and straight, along with hog rings and the pliers to clamp them. There were women that came for the large plastic syringe cases to store their knitting needles and crochet hooks in. For a time, they came for empty gelatin capsules for some kind of decorative crafts they were making.

Every spring when the weather warmed up and the school year was winding down, the students descended upon us to obtain pieces of latex rubber tubing in the larger sizes. They used it, somehow, as a water gun by filling it from the pressure of a tap, then squirting it at their targets. They claimed it was better than anything else in a water fight. I had to remember to order in some extra spools when the spring fever hit the schools.

The magnets for installation in the stomachs of cattle were in demand for science projects, demonstrations, and the never tiring toy for the kids to play with. An article in an automotive magazine made the claim that a magnet or two on the gas line leading to the carburetor would super-activate the molecular components of gasoline as it passed through the magnetic field. It was supposed to give more power and increased gas mileage to your car. We sold a lot of cattle magnets that didn't end up in the cows' stomach.

One day, Dan Baker came in for a needle three or four inches long. He was one of those people with a big broad grin who appeared to be eternally happy. His face wore a perpetual smile that was just waiting to gush forth in a bubbly laugh. I always greeted him as Dan, and he called me Dr. Dan in a slow-moving accent from somewhere in the South. He was a joy to be around and I cherished his friendship over the years. He worked in the oil patch and ran a few head of cattle on the side. My first impression of him was his tremendous strength. We were trying to get a heifer out of a truck and he simply took her neck in his arms and lifted her front half and dragged her out. It wasn't mean, rough, or in anger: just kind, but firm.

This day when he wanted a big syringe and a long needle, he explained that he was going to cook a turkey, and he wanted to inject his secret formula of salt and seasoning right into the meat of the bird. We had some sixteen gage three-and-one-half-inch needles we showed him.

"That's just what the doctor ordered," he said, as he laughed and explained how injecting the birds made them taste a whole lot better. He must have been right, for some of his neighbors and friends came in for them, or had him pick them up.

It was in what might be termed my twilight years when I had retired, at least from the responsibility of keeping the clinic going and the animals healthy. With the profession being kind of a twenty-four-hour, seven-days-a-week obligation, I had turned the business over to my son, Dr. Mark Dennis, and Dr. Blaine Whiting, and I was taking it a little easy. Since they kept the clinic open six days a week, it was a challange during the busy seasons, to give Arlinda, the office manager, a day off. Mark had asked me to come in and work on Tuesday while she was gone.

We were quite busy with several clients in the office being helped, when a stranger came through the door. He was a good-sized man with some dark hair showing through a lot of grey. His eyebrows were bushy and he wore a mustache, both of which showed prominent streaks of grey. He was followed by a younger fellow in his late teens or early twenties who showed a lot of resemblance. I wondered if it was a son or even a grandson. When the others had all been helped, I turned to them.

"Do you sell needles and syringes here?" the older man said. This was a request that alerted a warning signal, since the drug culture had emerged and laws had been passed that made it illegal to sell the paraphernalia used for illegal drugs. They were strangers, so I thought I'd better be careful.

"We sure do, what do you have in mind? We have the plastic disposable, nylon, and metal ones," I said, thinking if he were a livestockman he'd probably know just what he wanted and it would be legal to sell him some.

"I guess the plastic disposable kind would be the best," he said without any further explanation.

"What size would you like?" I asked, and noticing some indecision on his face I had a better idea. "Why don't you come on back here and take a look? They're all in this next room," I suggested as I led the way around the corner into the supply room. I showed them a three, a six, and a twelve ml sized syringe. They looked at each one and made the decision.

"I think we'll take this one," the older man said, holding up the twelve ml syringe. That relieved me to some extent, because most drug materials I had seen on television were smaller sizes.

"How about the needles?" I said as I pulled open the drawer filled with boxes of various sized needles. I pulled the cover from an eighteen gage, one-and-a-half-inch needle and held it up. This was the most popular size used for horses and I felt they must be getting it for a horse.

"Have you got anything smaller?" he said as he looked at it closer. I picked out a ninteen gage, one inch, the smallest we had in that drawer, and showed him that. He shook his head and my curiosity began to build.

"What are you going to give shots to?" I inquired, while watching close for any reaction that might surface while he answered. I was still suspicious since large animals seemed to be ruled out.

"My worms, you know, night crawlers," he said with a fleeting half smile on his face. "What kind of a kook is this?" I thought, and then reconsidered...Science and research had effectively passed me up and maybe they were treating worms. They could be giving them steroids or something to make them grow bigger...I better not show my ignorance. I guess the confusion showed on my face. "I'm going to shoot them full of air so they'll float," he said in explanation.

The realization swept over me, they were going fishing. I'd noticed the nice boat behind their camper-covered pickup when they pulled in.

"Oh! I see, that sounds like a good idea. I'll have to go into the small animal room to get you the needle you want," I said with a chuckle as I went to get them. I brought out a twenty and a twenty-five gage, and met them at the desk.

"That's the one," he said, as he saw the small bore twenty-five gage needle. "How much do you need for them?"

"Fifty-five cents," I said, as I added up the sum in my head. He handed me a dollar and as I dug out his change I asked,

"Where are you going fishing, I guess you're on your way, with the boat and all?"

"Lake Borum, they're getting some whoppers out of there. The other day I saw a trout a fellow caught off the dam that was this long." He held out his left arm and touched it halfway between the elbow and his shoulder. "And there's some good sized bass in there too." I could see they were getting excited to try out their worms.

"Well, you have a good time and catch a big one," I said as I handed them the change, and they were out the door. As I entered it in the daybook I wondered what Arlinda was going to think when she saw "worm" written in the space for animal species.

Restored Rooster

I returned to the clinic one afternoon and Dr. Whiting said, "We've got a patient for you."

"Where's it at?" I asked as I glanced first toward the small animal section and then toward the back. Most patients they consulted with me on were cows, and I hadn't seen any when I stopped by the corral.

"It's in one of the outside kennels. It's a rooster. It belongs to Etta McCurdie; the little, older Indian lady. She brought it in this morning."

"A rooster?" I asked with a puzzled stare.

"That's right. She brought it in and handed it to us and said, 'My rooster is sick. Maybe you can fix him up?'"

That sounded like Etta McCurdie all right. She was a sweet little Indian lady who had been bringing her dogs to me to board and vaccinate for years. She was short, not much over five feet tall in her moccasined feet, with skin that was weathered by time and hair with a touch of grey for as long as I could remember. She loved her dogs and always referred to them by name. At first there was Lady, then Furky. The two I remember best were small, furry crossbreds she called Noney and Winchey. She boarded them periodically at the clinic for many years. Whenever she would go out of town, she wanted them cared for. I remember how disgusted she was with her granddaughter when Etta had had to bring in her dog that needed attention. She had been neglecting it and Etta said, "She shouldn't have a dog if she doesn't take care of it."

I went out and looked at the old rooster. He was down on his breastbone with his legs twisted sideways. Most of his feathers were missing and his skin was blue. Overall he appeared to be knocking on death's door. I wondered if he was just an example of a flock problem from which we could make a diagnosis.

Over the years, I had been contacted many times for help with chickens, turkeys, pheasants and birds that were sick or having problems. The usual pattern was someone bringing in sick or dead birds for me to diagnose. Then I'd make recommendations for treatment of the flock. In this, I frequently needed help and often contacted Dr. Royal Bagley with the Moroni Turkey Cooperative. Sometimes I would send him samples and birds for laboratory diagnosis. He was one of the world's best poultry specialists and a special friend.

"What's the deal with the rooster?" I asked as I returned to the office.

"You're supposed to tell us what to treat him with," Mark said.

"Treat him! I think it's too late for that. Is she having a flock problem?" I asked as I reviewed the picture of him in my mind.

"Flock problem! That is her flock. He's her pet and she wants him fixed up," they explained.

The overall image in my mind indicated a possible malnourishment and the way he handled his legs conjured up a memory of baby chicks I'd seen over the years with riboflavin deficiency. I remembered giving them dried whey or milk products. There were other vitamin deficiencies that affected the nervous system, but I couldn't remember about them.

"I think you'd better give him some milk or cottage cheese. I think it could be a riboflavin deficiency, or one of the vitamins. I'd have to look it up," I said as I racked my brain to try to remember. Poultry diseases didn't come around often enough to have the answers on the front burner.

"From the looks of him, it sure could be a deficiency. Maybe more than one," Dr. Whiting offered. "What we should probably do is mix him up a little tonic with the vitamins in it." That was what they did. They gave him some milk and put the other vitamins in it. They fed him a little dog-and-cat food for protein along with a grain mixture. They gave him the liquid mixture from a syringe in his mouth at first, but soon he began to eat by himself.

In about three or four days, a miracle took place. He was able to stand on his feet and act like a normal chicken again. When Etta came in and saw him, she was just overjoyed. She laughed when he strutted around. I think she was a little surprised to see him alive, but I don't think anyone could have been more surprised than all of us at the clinic were. She paid her bill, gathered him up in her arms and smoothed down his feathers. As she went out the door, she was talking to him and calling him "My Captain."

13
INTO THE HILLS

High Mountain Surgery

Sometimes the lessons we learn by our mistakes stick with us longer than those we learn by success. They tend to haunt you for years to come. One such experience was a call I got from a man who wanted me to operate on a horse he used in the timber. It had a large sarcoid, or blood wart as he called it, on his foot. In the timber and brush it got bumped and torn and was always bleeding. I explained that my success rate wasn't too good, and that many of them came back. He was willing to try, because the horse wasn't any good there on the mountain the way he was. He wanted me to come up on top of Mosby Mountain and operate up there, rather than him taking time off to bring the horse down. It was the middle of the summer, and a trip to the mountains sounded like fun, so I made him a good deal on the mileage and we set the time for an afternoon.

When I got to the logging camp, it was a perfect day with bright sunshine and no wind. I'd eaten my lunch on the way up and arrived just as they were ready to start the afternoon work. They unharnessed the old horse and we picked out as smooth an area as we could find to lay him down on, and built a fire to heat some irons. I thought I'd do the operation like I had for Parley Lambert's horse, only I would give him an anesthetic to help get him down and minimize the pain. The only anesthetic of choice for the horse at that time was chloral hydrate, which lasted up to an hour and had a long, difficult recovery period. We would sit on the horse's head and try to keep it down just as long as possible because when they got up, they were very dizzy for quite a spell. They would stagger and fall many times if there wasn't help enough to keep them balanced on their legs.

When all was ready, I put a big needle into the jugular vein and ran in the anesthetic as rapidly as possible. He didn't stagger much before going down. I felt I wanted to keep him just a little on the lighter side, because it wouldn't take too long to cut the thing off and cauterize it with the hot irons. When he relaxed and showed that he was under, I shut off the anesthetic. He was a big horse and the sarcoid was rather extensive, covering a little more than half the foot below the fetlock. I put a rubber tubing tourniquet on his leg to control the hemorrhage, then went to work cutting and burning. When it was all peeled off, we loosened the tourniquet a little to check the bleeders and cauterized them, then I powdered it with caustic dressing powder and bandaged it up. We took off all the ropes and left him to recover enough to help to his feet. I put my ropes and equipment away and laid the hot irons out to cool and put out the fire. Then we just visited and I walked out around to see the timbering process for fifteen or twenty minutes. When I got back to the old horse and checked him, the world was still spinning for him. I thought he should be recovering, but wasn't worried, because he was breathing normal and groaning now and then. When another half hour had passed, my concern kicked into gear, and I began to think. I monitored the heart rate which seemed slow and checked his mucous membranes which were bluer than normal. What was happening to cause this? Why wasn't he detoxifying the anesthetic? Maybe it was my own shortness of breath that alerted me, but all at once I realized that it must be the altitude. We were right on top of the mountain and must be near ten thousand feet above sea level. The experiences I had in decompression chambers and flying told me, that old horse needed more oxygen. I asked them if they had a welder or cutting torch in camp and they did. We brought it over and began to bleed oxygen into his nose and sure enough we could see a difference. In about ten minutes, he began to rally and rolled up like he wanted to get up, but without the oxygen which we couldn't keep in his nose, he soon went back down. We held him and ran oxygen in his nose as long as we could, but he would throw up his head and make an effort and we'd loose ground again. I think we might have been successful if we hadn't run out of oxygen, but there wasn't much in the tank and within an hour it was gone. The owner finally said, "Just let him go, and he'll come around." I couldn't think of anything more I could do to help so I came back home. I never did hear what happened to the horse, but I learned a lesson that said to me, You didn't use your head and think things through. If you had thought about the altitude, and how much blood the horse had lost from hemorrhaging every day, you wouldn't have operated on him on the top of a 10,000 ft. mountain. It certainly took all the fun out of a trip to the mountains.

Meadow Creek Excursion

"Hello. This is Dr. Dennis." It was around eight in the evening and I was home with my family.

"Dr. Dan, how are you? This is Jim Bell." Jim was a veterinarian from the Cache Valley area. He graduated from school a few years after I did and had developed a reputation of being a wheeler-dealer type. He was a good veterinarian and had himself involved in some consulting roles with some large ranch operations. I wasn't fully aware of all that he had been or was involved in. I wondered if he was in town.

"Are you in town?" I was curious to know what his interest might be in this area.

"No, I'm home, but I'd like to come out there tomorrow. Do you know where Meadow Creek is? It's out in the Book Cliffs someplace. I ran into an operator from out that way awhile back and only got to talk to him for a minute. He said he ran around eight hundred cows and I am interested in selling him some protein block. You probably already know I'm selling block for that block company over in Cache Junction." My mind had been adding things up as he talked, and besides selling protein block out in Meadow Creek, I could see an interest in pregnancy testing and checking bulls.

"I've never been there, but I know where it is, and I could give you some direction. It's quite a ways out, probably sixty or seventy miles the other side of the Green River and it's all dirt road." I didn't want him to think it was a short trip.

"What are you doing tomorrow? I wondered if you could go along and we could have a good visit?" Jim was a good friend but I was a little suspicious of his wanting to talk to me or come into my area. He had a tendency to portray himself as the out-of-town expert, and he did have the ability to persuade cattle operators of the benefits of herd management, pregnancy testing, and checking bulls for semen quality. I knew he had some larger herds over the mountain in Wyoming using his help, at least in the fall. His pattern was to line up the larger herds for a couple of days in the fall when he could come in, "skim off the cream" as it were, and leave the local practitioner with the more mundane and less lucrative work. In a sense, these out-of-town experts were parasitic to the local practitioners. In the small animal arena, they had spay and neuter clinics, sometimes in mobile facilities that could blanket an area. I wasn't anxious to do all the night and weekend calls on a twenty-four hour basis, and then have some golden tongued expert come in and rob me of more lucrative opportunties.

"I don't have anything on the agenda for tomorrow right now, I guess I could ride along. I love to go out in that area." I thought it would be fun, to see some new country and forget it all for a day. I wasn't too worried about him and the ranches out that way, I knew how they operated.

"Good. I'm glad you can go along. It'll make it a lot more interesting day, and with your knowing the lay of the land, I probably won't get lost. I'll see you in the morning." His enthusiasm was catching. I was looking forward to it, even though it was a long hard trip.

"You better get on your horse before the crow pees in the morning and get out here as early as you can. That's an all day trip." I didn't want him to get here at noon and expect to get out there and back.

"Sounds like it. I'll try to leave a little after four. I should be there around nine. That sound all right?" I had to admire his willingness to get up and get going.

"That'll be fine. I'll see you in the morning."

When he arrived the next morning, it was in a small or medium-sized car. I took one look at those small tires and concluded that if we were going to Meadow Creek, it would have to be in my truck. Maybe he had planned it that way.

"Jim I think we better take my truck. I don't think that thing is high enough to make it out there and back." I thought if I was going, I didn't want to get stranded out in that country by a sharp rock going through an oil pan.

"Is that right! Well it's all right with me, if you don't mind." He was eager to get his stuff out of the car and into my truck. I hadn't planned any lunch, because I didn't know what he might bring. It didn't look like he brought anything. I did have plenty of water in my little tank. We topped the gas tank off at the gas station and he paid for it. Then I pulled in at the grocery store.

"I think we better get us some groceries. It's going to be a long day." I picked up a package of cinnamon rolls and a package of bologna, my favorite sandwich material, then four cans of pop and some Fig Newtons. He got some vienna sausage, apples, candy bars and Twinkies. At the checkout counter, we threw in a package of licorice, and we were on our way. I pushed right along as fast as I dared on the pavement and we crossed the bridge at Ouray about ten thirty. We shared some philosophy on practice, politics and things in general as we went along. I kept him informed about the area and shared some history of Fort Duchesne, Randlett and Ouray. It really was an enjoyable interlude to my everyday routine.

When we crossed the White River and hit out across those bare hills I think he began to realize why I wanted to bring the truck. I told him about my Uncle Hank and his ferry boat and showed him where the road turned off to his ranch located on the lower part of Willow Creek. I think he was wondering how a cow could even live in a barren place like that. When we came to the forks in the road, I explained how one road went up the ridge while the other went down into the creek bottom and up both Hill and Willow Creeks. I related the story of how the Turkey Trail Dugway got its name, and about my other Uncle Ephriam Birchell's ranch at the mouth of Hill Creek. Hank and Eph were really my mother's uncles, one on each side of her family, and my life had been enriched by many tales about their lives. He could understand that I had some roots in that part of creation.

We went up the ridge road where we saw some rabbits and prairie dogs scamper for cover as we went by. There were a lot of crows eating the carcasses of the ones that didn't make it out of the way. When we came to the edge of the high ridge looking down into Willow Creek, I stopped. We answered nature's call and walked out to the edge where I pointed out Hill Creek and where the Green River cuts down through. I think he was a bit awestruck by the immensity of what we could see and the courage and strength of those who first came there added to it. I know I was. We hurried on up and down the roller-coaster road through the cedars to the top of Buck Canyon. I told him I wanted to go down and talk to Willis

Stevens about the road to Meadow Creek. I was sure he or one of his men would be at the ranch. We had seen a couple of deer coming up the ridge, but when we turned down Buck Canyon, I think we saw upwards of twenty head. I couldn't look at all of them and keep the truck on the road. When we got to the bottom, I took a left up the canyon and we were soon pulling into Willis's place. He and his men were coming out of the house as we drove up. After having looked down into Willow Creek from the ridge, Jim had asked me about the ranches and I could tell that his interest had picked up.

Willis and I were good friends and had been for years. I'd met him up at the Blank's place in Neola. I think Mrs. Blank and Willis were brother and sister. Anyway, they were related, and he used to bring cows to Neola. I had taken a cancer eye or two out for him over the years. He was tickled to see me and gave us a hearty welcome. I was glad we'd caught him at the house. I had a map and knew roughly where we were going, but I didn't know much about the roads. I was expecting to go back up Buck Canyon and around by P R Springs. I introduced Jim and explained that he was selling protein block. Jim was never bashful and immediately began to question Willis about how many cows he had, when they were corralled in the fall, how much he paid for protein and the whole bit. I knew Willis only corralled his cows in the range corrals to pull the calves off and never worked them all at once. He soon told Jim that he preferred the protein pellets, because he felt the cows got a more equal chance at them, and sometimes it was the only way he could get them in a corral. He didn't have much use for the block because they got wet, then froze, and the cows broke their teeth on them.

I could see Jim was let down a little and it was time to move on, so I asked Willis about getting to Meadow Creek. He said we could go on up the ridge, but said the shortest way was on up the canyon. He looked up at the clouds. It was overcast and looked like it could rain. He explained that up the creek just before we got to the Poweenie Ranch there was a road that went up a narrow side canyon. He said it was boxed in on both sides and only wide enough for a truck. He said it wasn't very long, maybe a mile and had a rock bottom all the way. Once out of the canyon, he said there was a road up a long sagebrush slope to the other road. Take a right there and the next left would be Meadow Creek.

"You better hope it doesn't rain, because that box canyon can flood, and that road up the slope can be a muddy son of a gun," he told us. I could read the warning in his voice.

"Thanks, Willis. From the looks of those clouds, maybe we'd better hurry. I'll see you around." With that we hurried to the truck and started on up the canyon. We ate some lunch as we negotiated our way along the winding road and through the gates. It started to sprinkle and I wondered if we should turn around and go on the other road. I knew how suddenly it could turn to a downpour. It kept sprinkling very lightly and I pressed on. When we came to what Willis had described as the side canyon or draw I wasn't sure we'd found the right one until I saw some tracks leading over to it. It was raining lightly by this time, coming steadily without wind or lightning, so I took courage. The floor of that draw was solid rock and water was starting to run down it. The walls were so close we could almost touch them, but it was only supposed to be a mile long. I drove on through the water and when we were what I'd guess was a third of the way, the rain began to fall much faster. They weren't big drops like a cloud burst, but I wondered if I should try to back down out of there. There hadn't been a word spoken since we entered the narrow gap so I didn't say anything. I finally concluded I could get out the top about as fast as I could back down, and I couldn't turn around. I increased my speed as much as I dared and the water would sure spray out from the tires as we pushed up the stream. I had a few visions of meeting a flood around some corner, however it didn't happen, and the water gradually increased to where I had to go slower. The rain stopped suddenly when we were about a quarter of a mile from the top and our spirits lifted materially. By the time we reached the top, the little stream was about six inches deep and three feet wide where we crisscrossed it a few times in the last hundred yards. The climb out was up over little stair-stepped ledge rocks each a few inches thick, with the water spilling down over them. We slipped and spun a little as we climbed out on to a solid flat rock table a rod or so around the top. I pulled to a high spot and stopped. We breathed a sigh of relief and got out.

"That got to be a little hairy before we got out of there," Jim said smiling.

"I'll say! I was beginning to think we were between a rock and a hard place, and time wasn't on our side." (That wasn't meant to be a pun, but we sure had been in between the rocks.) "Maybe we still are, from the looks of that road ahead. I think we better put the chains on here where it's clean," I said as I looked at those tracks leading off through the low sagebrush up that long slope. It looked at least five miles or more to the top of that ridge and it was uphill all the way. We ate a Twinkie apiece and opened a pop to wash it down. Then we put the chains on and he cinched them up as I rocked the truck forward and back. When we figured they were good and tight, we made a quick rest stop and headed out into the mud. The rain had only soaked down an inch or so, but that clay was as slick as a greased door knob. The chains would dig down to dry ground and send us forward with each revolution. The mud rolled up around the front tires like making a snowman until it hit the fenders and springs or broke off on the outside. It was slow going and the mud became more thickened and stuck to the wheels more as we went along. Our tires appeared about fifteen inches wide as that mud would roll up on them. It took about an hour to make the five miles, but the sun was shining and it was good to smell the rain soaked sage. The road on top was a graded road and fairly dry. I pulled on to it and headed into the sun which was getting down near the horizon. I speeded up for a half mile to spin off as much mud as I could, then we stopped and took the chains off.

We clipped off the miles at a good rate for a dirt road and were soon up to the Meadow Creek turn off. There was a rather steep dugway with a turn partway down, and I could see Jim tense up as I headed down, but I could tell it was shale and nothing to worry over. The ranch buildings were a short way up the draw and we arrived about the time the sun disappeared for the day. It had been a tough trip, but I had delivered him where he wanted to go, and I felt good.

Jim made his sales pitch on the block, although he didn't say anything about delivering it like he had to Willis. He didn't ask about preg testing or checking bulls and it was soon obvious they were not ready to order any block. They asked us if we wouldn't stay for supper, said the wife had a roast cooked and it was all ready. Jim took a furtive look at where the sun had gone down and told them we couldn't stay. It would have been a real pleasure for me to stay and find more out about the ranch and the area, but I could tell Jim was plenty worried about that dugway and the trip home. I had already made up my mind to go back by way of PR Springs, although I hadn't shared that with him. I guess he could see that muddy five miles and narrow draw in the dark. At any rate, we thanked them, got in the truck and made for the dugway. It was a breeze and we relaxed and ate some more of our food as we headed down the road. When we came to our muddy tracks, I didn't turn off and kept on going on the other road. Jim turned to me with a questioning look and I explained about going home the other way.

"I hope it's this good all the way," Jim said as he slid forward in the seat and leaned back. I could tell he was tired after getting up so early. Down the road about five more miles two black bear lumbered up on the road into my headlights and scooted off across the road in front of us. We commented about the rarity of seeing bear and how fun it was. Jim was soon asleep. I reflected about the trip and the area on the four-hour drive back home. To me it was a memorable day. To Dr. Jim Bell I think it was a little disappointing and scary at times, but I think he gained some appreciation for the challenges of the basin.

Castration Day at the Chew Ranch

Doug Chew had talked to me at a cattleman's meeting

about castrating some horses for him. We agreed upon a day and when it came, I proceeded to Jensen, crossed the Green River and followed the highway a few miles out on the plateau. I found the one-track road leading north as he had described it to me. Driving north along the west end of Blue Mountain and directly toward Split Mountain Gorge gave me a lot to look at. I had never been that close to those mountains before. Trying to see the solid rock formations and extremely rugged escarpments carved by the water over eons and stay on the road kept me busy and fascinated too. I drove in awe-inspired silence until I came to the trickle of water called Cockle Burr Creek. It was running over a solid rock surface where the road crossed it. The crossing reminded me of a story told of an earlier time.

The time setting I'm not sure of: probably in the nineteen twenties. At any rate Joe Haslem, a very tall and slender cowboy from Jensen, was called upon to take a bureaucrat from Washington back into the area. They left Jensen by buggy and by the time they arrived at the crossing of Cockle Burr Creek they were thirsty enough for a drink. Joe stopped the horses short of the creek and they got out and approached the water. The bureaucrat was careful to proceed to the area above the crossing and Joe went to a pool below. When he kneeled down to drink, the bureaucrat cautioned him about the possible contamination below the point where the animals crossed. I guess the fact that cattle, sheep and wild animals used the stream above there didn't cross his mind. Anyway, Joe is reported to have said in his typical western drawl, "I don't know's it makes a hell of a lot of difference when you plan to drink the whole thing anyway, does it?" He then lay down and had his drink.

As I got further north, approaching the Green River, I was still on a little plateau and the road turned east. There was a narrow deep ravine coming in from the south and the road went right to the edge of it and turned down the top of a tilted ledge. The rock formations were all tilted on a ten- to fifteen-degree slope and the road went down the edge of the ravine on a narrow ledge just wide enough for a one-way road. When it came to the bottom, it crossed the sandy wash in a sharp turn and climbed back out on the corresponding ledge on the east side. It kept your attention as you negotiated the ledges, but it was a unique way of crossing the ravine, probably the only way. It wasn't far until the road dropped down to the Chew Ranch nestled on the bottomland of a big sweeping turn in the river below Split Mountain. The setting and the scenery were more spectacular than a story book setting and difficult to adequately describe. However, my greater treat was yet to come as I watched a master at his trade.

I followed Doug and his two cowboys down country about a half mile to a round pole corral with seven or eight horses in it. I asked Doug if they were broke. He told me they weren't. They were all two- and three-year-old stallions that had never had a rope on them. My heart said, "Oh no, what a job this is going to be to rope and throw each one of them. This is going to be a long day." I got out my emasculators and put them in my coverall pocket, put the scalpel with a new blade between my teeth and was about to pull out my throwing rope when I noticed they had all climbed over the fence with a lariat in their hand. They hadn't asked me for any direction so I walked over to the fence to see what they had in mind. The corral was oval in shape and Doug was standing in the middle. The two cowboys were at one end with the horses, attempting to string them out one or two at a time as they hazed them past Doug. He was standing with a loop in his rope held partly behind him. As the last horse went by, he flicked his arm forward. That rope sailed out and the loop seemed to stand upright and still for a second or two while the front feet of the horse went through it. He jerked the rope tight and set his weight against the rope with his hip as the horse moved forward and his feet came off the ground. The tight rope doubled the front legs back and the horse crashed to the ground. One cowboy rushed up to help Doug keep the rope tight on the feet while the other one circled around and got on the horse's head. As soon as he had a good grip on the head, the first man put his rope on the top hind foot and pulled it forward and up to the top of the fence and dallied it. That was my cue, I hopped over the fence, stood in behind the horse, reached forward and did the surgery. I stayed up toward the tail so the free hind leg couldn't get at me and my knee against the butt of the horse kept me balanced. When I finished, the cowboy with the leg rope took it off his leg and we all stepped back between the horse and the herd as he got to his feet, letting Doug's rope fall to the ground. One cowboy opened the gate, and we hazed him out through it. It was all over in a matter of minutes. The process was repeated with me standing back against the fence each time to marvel as I watched the master at his work. The most difficult chore was getting the horses separated to go by Doug one at a time. We finished in a little more than an hour. I think they appreciated my skill as well because as I was washing up, one of them made the comment, "You're a good man, Doc." As I made my way back over the prairie I thought what a treat the day had been. I wanted to come back again sometime in the future.

Cancer Eyes at the Green River

"Doc, could you come down to my place on the Green River and take some eyes out for me?" It was Ray Sprouse. I could tell his voice, however, I wasn't sure what place he was talking about on the Green River bottoms.

"How about tomorrow morning about nine-thirty or ten? That be all right for you?" I could get everything ready and be there by that time.

"We'll be all ready for you. Do you know where to come?" I was glad he had asked because I didn't know.

"Not for sure, Ray. You'd better walk me through it so I'll know the right road." I knew three or four roads that went down to the river.

"You know when you're going straight south past Pelican Lake, you keep going until the road turns to go to Ouray. Take a left on that road there at the turn. Keep going east on it and it'll bring you to an old house with four big cottonwood trees just south of it. That's the place and we'll be there watching for you." I visualized what he told me.

"I've got it. I'll see you in the morning. By the way how many do you have?" I thought I'd better know how many to prepare for.

"I think there's six of them, Doc, if we can find all of 'em. They get hid up in the willows and tamarack and it's hard to find them sometimes. Anyway we'll do all we can find. See you in the morning."

The next morning I put in plenty of Novocaine, filled my antiseptic scrub bottle, checked my kit for extra scalpel blades and filled six jugs with warm water. I filled the water tank on my truck and took off. When I drove up to the old house, Ray's truck was there and he and his hired man were on their horses. They had seen me coming and climbed aboard their mounts. I pulled around the old house and expected them to lead me off to some corrals. He hadn't said anything about corrals and a chute when he called, but I just assumed he had some. I wasn't expecting any power to be available so I'd have to clip them with scissors, which I'd done many times before. I stayed in the truck and Ray rode around to my side.

"Where we going to do them?" I asked, expecting him to tell me to follow.

"Right there between those four trees. We'll go bring them in and I figured we could stretch them out between the trees." With that he reined his horse around and spurred him into a gallop off toward the river with his help right behind. The realization of what he meant left me in a little shock. I'd never attempted to remove a cow's eye except with its head tied solid in a squeeze chute. I suddenly realized that Ray was of the old school of cowboys who figured if it couldn't be done on the end of a long rope, it couldn't be done, and we were going to do it his way. It was going to be a new experience. There wasn't much I could do to prepare other than to set things out in readiness in the back of the truck. It wasn't many minutes until I saw Ray coming with his rope around the long sharp horns of a big Hereford cow. His hired man was behind with his rope on a hind leg. Ray came around in between the trees. She wasn't hard to pull, because every now and then she would make a dive for his horse. His horse had to be quick and if that rope hadn't been on her back leg, she would have enjoyed ripping into that horse's guts with one or both of her long sharp horns. She was wild and mean and wanted everybody to know it. The hired man stopped his horse and held the hind leg while Ray stretched the mean old gal out. I took my cue about then and grabbed hold of her tail. With her foot stretched out behind, I was able to pull sideways on her tail and throw her down. I grabbed her front foot as she floundered and pulled it back, kneeling against her ribs with my knees. Ray transferred his rope, which was on the horns, to the tree and tied it. He and I held her down while the hired man loosened his loop and put it on both hind legs. He could now sit on his horse and stretch her out till she about came off the ground.

Ray wired one end of a pole down low on one of the side trees and pulled it around and across her neck. He could hold her head down with that. It was all up to me now. The bad eye was up and the cow was under control. I got my kit and a jug of water from the truck, knelt down by her head and went to work. I clipped the hair off and scrubbed and disinfected the area. As I started sticking the needles into the skin and behind the eye to deaden it, she tried to move, but Ray sat a little harder on the pole and she gave up. It was a little different with the cow on her side instead of upright, the eye socket was full of blood and I had to operate by feel instead of sight. I was surprised how easy and how fast it all went. I sutured up the opening, put my instruments in the kit and took them to the truck.

When I returned, Ray had the pole off her neck and was on his horse, which was by the rope on her head. I untied the rope from the tree and let him dally it to his saddle horn. Now all we had to do was get the ropes off her. I stood more or less on her lower horn, with the top one for balance, as I got a good hold on the rope to pull it off. I nodded to Ray and he gave me plenty of slack. I stepped back, pulling the rope off the horns and headed around a tree. At the same time the hired hand rode forward, loosening his rope, and the cow jumped to her feet allowing his rope to fall to the ground. I went around the tree and headed for the truck and that old sister came in hot pursuit. The two horseman had spurred out of range, so I was on my own. I rounded the back of the truck as she came down the driver's side and I made it in time to get in the passenger side and slam the door. She kept me imprisoned in that truck by blowing snot, shaking her head and pawing the earth right outside of the doors. I reached over and rolled up the window on the driver's side. I was afraid she was going to try to dive through it. I decided to drive the truck to get away from her. When I started the engine, she turned tail and headed for the river. I think they were the meanest cows I ever worked on. They found three more and each worked out about the same, except that we had to roll two of them over after they were down to get the bad eye up.

Ray brought the other two cows up to the clinic later, and they were just as mean as those on the river. They would charge the fence and turn their heads to get a long horn between the poles where our feet were. We'd have to balance on the top of the fence to get our legs back out of the way. I began to realize why Ray didn't handle them in corrals much. They could kill a man or a horse in a pen if they couldn't escape quickly enough.

Flying Vet

When a local group of pilots got together to form a flying club and buy an airplane, I joined them. We started with a four place Cesna 172, then in later years we bought a 182, which had more power and speed. I loved to fly, but time was precious in trying to keep the practice afloat, so I didn't do a lot of it. The plane did make it easier for me to attend some of the vet meetings and continuing education seminars, since I could go and come back in a few hours, instead of taking more time off and killing myself driving all night. It also provided me with the opportunity to help out some cattle owners whom I took out to search the mountains for lost cattle. Once I was called upon by the wildlife officers of the Ute tribe to go out and count the antelope herd on the other side of the Green River. I didn't use it a lot in my practice, but one time in early winter a group of cattlemen from Daggett County wanted me to come up and bangs vaccinate their calves. To save driving time, I told them to meet me at the Manila Airport at eight-thirty and I'd come by plane. I took a man along to help me and we were there as scheduled. I could see the strip was blanketed in snow, but I was surprised when I landed to see that it was about five inches or more deep. It took considerable power to taxi over to where they were parked waiting for us. We vaccinated and doctored cattle all day, and finally, when the sun was getting low, we finished our last bunch. When we got back to the plane it was considerably colder than in the morning, but the plane's engine turned over a few times and sprang into life. I let the engine warm up while I stowed things away and checked things out. When I waved to the ranchers and started to taxi, I found that snow was a lot firmer and hard to move in. It worried me a little. I knew the strip was about seven thousand feet high or more and a plane couldn't develop its full horsepower at that altitude. I did my pre-flight check and lined up with the runway, easing the throttle forward to full power. The plane was sluggish and didn't pick up much speed and when I hit my tracks from the morning landing, it slowed me some. I steered to the edge of the strip away from my tracks and kept going. I was halfway down the runway, and I knew I had to make a decision to abort the takeoff or keep going. The strip was a little downhill the way we were going and there were no fences or trees at the end, with a valley that fell away I could turn down into to pick up speed. I decided to go for it, although I was past where I should have been airborne. The speed gradually increased and I eased back on the controls and we broke loose from the snow about a hundred feet from the end of the runway. The plane felt mushy and I eased the nose down into the valley and breathed a sigh of relief as we picked up speed and began to climb. Suddenly I felt tired after a long day's work.

Another time, a riding club from over at Price arranged for me to come over there to castrate some horses. I told them I'd fly, and to pick me up at the airport. I went from Roosevelt to Price, castrated three horses, floated several horses' teeth and was back home in half a day. It was about like a call to Mt. Home, and to think I was an out-of-town specialist, since there weren't any vets in Price then.

When I was the only vet from Craig, Colorado, to Salt Lake City, I was called upon to make some long calls. One of these was out toward Piance Creek south of Rangely, Colorado, and I used the airplane for it. The Hill family, who lived in Vernal, had a ranch on Piance Creek, and they wanted their calves vaccinated for brucellosis. I told them to meet me at the Rangely Airport and I'd come by plane. They said I could land in a pasture at the ranch, but we had a gentleman's agreement that we would only land our plane at airports, so they met me at Rangely. I vaccinated over two hundred calves, did some cancer eyes and was home before dark. It sure did cut the travel time, and I thought many times how convenient a helicopter would be to practice with here in this scattered area. I probably wouldn't have been able to afford gas for one, if someone had given it to me.

Long Distance Racing

Between about the third and the sixth years after I came to the basin to practice, we had a rather active booster club organized called the Bull Berry Boys Booster Club. They got interested in long-distance racing as a means of boosting attendance at their Labor Day rodeo. There was much discussion about man being able to outdistance the horse on a long race, and eventually they decided to sponsor a race. They felt that if it was well-advertised, it could attract attention to the area and bring in some extra spectators to spend a few bucks and attend their rodeo. The interest mushroomed and the first race was scheduled. It was to be from Salt Lake City to Roosevelt and ended up with four entries: three horses and one man.

The booster club had a float decorated with bullberry bushes and sagebrush, and a band that rode in it played some lively tunes as it went along. They took this with some other entries, the horses, the human runner, and Harvey Natches in native Indian costume, and led a little parade down State Street in Salt Lake City. They had a good friend on the police force in the big city who was very interested in the race, and he helped them out with the big start. I'm not sure, but I think they started the race at twenty-first south, then from there to Orem, up Provo Canyon to Heber and out highway 40 to Roosevelt. The traffic turned out to be a bit of a headache until they reached Orem but from there on it moved along well. The horses soon outdistanced the man until that phase of the race was decided early, but he didn't give up. He ran all the way to Duchesne, and the club picked him up there to bring him in a hero, just the same. Mr. Abe Hatch, an older hardened rancher from out on Willow Creek, won the race and the other two horses crossed the finish line.

That first race did create a lot of interest and the next year they sponsored another one they billed as the Labor Day Trailride. Because of the traffic problem, this race was scheduled from the summit of Parleys Canyon to Roosevelt and had only horses this time. Whenever those who fancied horses or used them got together, it was the topic of conversation and much verbal wrangling went on. It had an appeal that a lot of people got caught up in. For instance, they debated which breed of horse would be the best. There were those who favored the Arabian, others the thoroughbred. Some were sure the Morgan was it while many put their trust in the quarter horse. Tales of speed and stamina from the past were spun every time friends would meet. I guess because I was a veterinarian and should know all the answers, I was quizzed and questioned on every turn, and the experts were quick to inform me with authority.

The more I heard, the more I thought about some advice old Doctor Swalberg from Utah County had given to my friend and classmate, Dr. Clair Porter, when he first started to practice. He said, "You want to be careful of racehorse people. Their words are no-good, their checks are no-good, their horses are no-good. They're just no damn good." That hadn't necessarily been my experience, but I could see a measure of truth in all the discussion that was going on.

Since we were the center of the activity, we had our share of contestants. Horses were started in training in every part of the basin. They figured they had to get them in shape and hardened into long hours and slower speeds. There were about as many ways to train and win this race as there were horses, and I wasn't going to favor one over the other. I never did know how many that started in training and never made the race, but I did see one die with azoturia in the training process. Like the draft horses of ages past, this one was kept on his full-rich diet over the weekend without the exercise, and then on Monday morning was started on his laps around the Neola track. His muscles tied up and the owner made the mistake of walking him a mile to get him home. If he had tied the horse up and not moved him, the chances are he would not have died.

The club set the purse at $3,000 for first place, $1,000 for second place, $750.00 for third, and $250.00 for fourth. It wasn't big money, but it attracted fifty-seven entries and most of them were there at Skyline for the big start. I was engaged as the official veterinarian for the race and given about the full responsibility for any disqualifications for abuse or neglect. In reality, I was more of a figurehead to give the race a little more credibility in the public's eye. At any rate, I determined to do all I could to make it a success and agreed to donate my time and expenses. As it turned out, it was one of the great learning experiences of my life and one which taught me an appreciation for horses that I didn't have before. I learned that horses would keep going beyond their capacity to endure and, like men on the battlefield, give their full measure of devotion by sacrificing their lives. When I began to check the horses that dropped out, I found physiologically that they were extremely dehydrated with evidence of muscle breakdown. I found almost no urine at all in the bladders, and what was there was as thick as syrup and dark with what I assumed was myoglobin from ruptured muscle cells. They stopped sweating for a time before other symptoms showed and ended up with tied-up muscles and irritated bladders that kept them trying to urinate. It didn't happen like the typical muscle tie up that occurs when exercise first starts, but came on them when they had expended all their available energy. From my observation, they were experiencing lactic acid buildup that causes the cells to go into spasmodic contraction primarily because of insufficient liquid in the body. The burden of responsibility pressed down very heavily, as I observed the number of deaths that occurred during the race and heard the reports of those that died in the next few days at their homes.

I wasn't able to know how each horse was watered and fed during the race, but I felt the common thinking of a lot of the short sprint racers that it is better to hold them off water before a race, may have contributed to the problem. I preached the philosophy of giving water frequently and all they wanted, together with feeding sometime during the race and thought it would solve the problem, but it didn't. The next two races, first from Duchesne to Heber over Wolf Creek pass, and from Heber to Roosevelt on the highway were better, but horses were lost on both of these.

In the 1958 race from Parley's summit to Roosevelt, Mrs. Dorothy Luck riding Cortez, a thoroughbred horse she got from Theron Horrocks, won the race by only a few lengths in a neck-and-neck-sprint-type finish down Roosevelt's main street in the middle of the night. The second place winner was a big bay horse from the Blanthorne Ranch out in Grouse Creek. He had been leading all the way until the turn at the south end of town. From there to the light at Lagoon Street, it was a close race. I don't think I have ever witnessed a display of stamina and all-out effort to equal what I saw that night. The old bay horse was so exhausted that his hind legs would buckle a little and he would drag his feet along the pavement, throwing sparks like an emery wheel until his shoes wore through. When I checked him after the race, he had ground off his hoofs until they were bleeding. The sorrel that Dorothy was riding gradually pulled ahead, with his gait so stiff-legged and stilted I was afraid he was going to topple forward on his head and injure her, but neither horse gave up until the riders pulled them to a stop.

In the 1960 race from Heber to Roosevelt, the same two horses came in first and second, only the big bay ridden by Mr. Blanthorne's son was first. Cortez fell down at the cemetery, but got up and walked across the finish line. He finished the race, but he died the following day at the Horrock's place.

The Bull Berry Boys did an outstanding job of boosting the community and the area, but I was glad they didn't have any more long-distance races. It would be my wish that never again would there be any long-distance horse racing, because owners, riders, and myself included are not able to tell when the horse has reached its limit.

14
WOOLIES AND PORKERS

Flocks of Sheep

When I came back to the basin to practice, there were a lot of small-farm flocks of sheep, and I would occasionally be called out on their problems. Generally it would be someone experiencing some death loss, who wanted to know what the reason was. I would go out and necropsy some of their dead ones to determine the problem, then tell them what to do with the herd to stop the losses. Frequently it involved parasitism and I could show them the inside of a stomach with worms thick enough to be like hair growing there, or a liver almost completely destroyed by liver fluke. Sometimes with a bad fluke problem, a big percentage of the herd would have bottle jaw, or a swelling under the jaw from circulatory impairment.

Sometimes the damage already inflicted by the flukes was severe enough that many died even after being treated, but it gave me an opportunity to recommend routine parasite treatment programs that prevented the situation in the future.

Another problem, not infrequently seen, was Black's disease and other clostridium infections such as red water, enterotoxemia and tetanus. The Black's disease was characterized by very rapid death losses occurring in twelve hours or less. Many times, I would find the animals lying in their normal resting position, as if they had gone to sleep and forgot to wake up. The inflammation under the skin along the neck and involving the front shoulders, together with the sanguinous fluid in the pericardium, would confirm my diagnosis. I soon found that in addition to vaccinating, a recommendation to move them to another pasture, especially if the one they were in involved a pond, stopped their immediate losses.

The finding of pneumonia in these small flocks, especially in the summer, was a surprise to me until I realized how gregarious they were and how closely together they bedded down. I guess the pathogen level built up and spread easily between them. I also felt the changing of pastures creating emphysema reactions predisposed them also, but an immunization program with a pasteurella bacterin helped a lot in stopping and preventing the pneumonia.

One of the most persistent misconceptions that I struggled with was the lamb losses the owners attributed to eating dirt. The real cause of death was enterotoxemia from the clostridium perfringens organism. The biggest, healthiest lambs, especially singles on a ewe, would die rather suddenly. When we got the owners to immunize the ewe while pregnant and then do the lambs at birth and again two weeks later, the losses were about eliminated.

I was called upon to treat a few individual cases such as delivering a lamb in a difficult birth or replacing and suturing a vaginal or uterine prolapse, or treating lambs with entropion of the eye, but mostly it was a flock problem.

As the time for lambing got close, I would have some calls

about pregnancy toxemia or pregnancy disease. This came about when the demand for carbohydrates to nourish the growing lambs in the uterus was not met in the diet. It could happen when the flock was thin and on marginal rations, or many times when the ewes were overly fat. If it was a flock that was fat, I'd always remember a statement made by my nutrition professor. He would emphasize the importance of adequate carbohydrate by saying, "Fat burns in the flame of carbohydrate." It was a good analogy. Because there wasn't enough carbohydrate to metabolize the fat, the livers became infiltrated and the light-colored fatty-infiltrated liver was the diagnostic clincher when I necropsied a ewe. The disease wasn't very successfully treated, especially if the animal was far enough along to be down, but adding a quarter pound of corn or other grain to the ration, or some other source of carbohydrate, would usually stop any more from getting it. I also felt it very important on the fat flocks to be sure they got some exercise every day. I'd have them walk the herd about a mile a day.

As I mentioned, sheep problems were most frequently solved by doing necropsies, and many times the odors were a little on the ripe side. One day Aaron Stevenson from Mountain Home, affectionately referred to as Swearing Aaron, brought two dead sheep down to me in the little jeep that he drove. It was a cold time of year and the jeep was all closed up and there wasn't much room inside. I knew the smell must have been very strong and mentioned that it must have been a long ride in there with them. He laughed and said, "You know a blankety-blank sheep is half rotten before they die."

McKeachnie Disaster

One spring I got a call from the McKeachnie family in Vernal. They had a large range herd out southeast of Gusher. They said about two hundred of their sheep had died or were down in one night. They asked if I could come and see what was killing them. My first thought was some plant poisoning, and the more I mulled it over as I drove there, the more it seemed the only possibility. There were dead sheep everywhere when I arrived. I quizzed them a little about where the sheep had fed the day before as I prepared to open some of them up. As I opened up a ewe or two, I was amazed to find the fatty infiltrated livers of pregnancy toxemia, and that was about all. I walked out through the herd, and sure enough, there were a few that seemed to be dreaming as they walked off alone. My mind clicked and told me it was pregnancy disease, but why? How come so many all at once without any on previous days. It was a real puzzle to me. I asked them if they were feeding any corn to them, and the answer was, no. Then the story began to unfold. Two days before, the herd was in two bands of about two thousand each. The owner had instructed the herders to move them in off the range to this location where they would join up for feeding and moving. The range was poor, so he had been hauling a little hay out to supplement them. There had been a death in the family and he was called out of town, so the sheep had not been fed hay for two days. That in itself probably wouldn't have mattered if it hadn't been the two days the sheep were moved. The combination of reduced grazing and the extra exertion of trailing the sheep pushed them over the brink. They were already on marginal rations and deficient in carbohydrate and the extra demand for muscular activity pushed them into the toxic stage. I recommended the addition of some corn to the diet and they increased the hay for a day or two and had no more losses.

The Big Puzzle

The Montwell Ward had a small flock of top quality sheep as a welfare project. One spring I was called about a ewe that had aborted, which puzzled the man in charge. He brought the dead lambs in for me to look at and necropsy. One was fairly normal, even though dead when born. The other lamb was completely necrotic or rotten. I was puzzled myself. I had never encountered anything like this before. I had seen vibrionic abortions in some herds, but this was different from that. I didn't have a good answer for him. In a day or two another ewe aborted, this time one lamb was born alive to die shortly after being born, and the other one macerated and dead as before. I went up to the farm, checked the feeding program, looked the herd over and checked the two ewes. They were eating and, like the others, in very good flesh and active. I came away without a clue, but I brought the two dead lambs. I wondered if anyone else was experiencing anything similar so I called the county extension agent to see if he had received any reports. He hadn't heard of any, but he told me there was someone in the area who would be going back to Logan that evening, if I wanted to send anything to the diagnostic laboratory. I packed the lambs in ice packs and sent them up to the lab.

The veterinarians in the diagnostic laboratory ran some cultures and tests and in the end they were as puzzled as I was. We consulted quite extensively on the phone and they wanted to come down and get some cultures from the ewes along with some blood samples. Two days later they arrived, and that morning another ewe had given birth to two lambs, one necrotic and dead, the other alive but only for a short time. We packed the lambs in ice, got the blood samples they wanted and they rushed back to Logan to do the diagnostic work. By the time the test results had all been read, they had researched all the literature on sheep abortions, contacted the state veterinarians office, the federal animal disease supervisor and called the national sheep research station at Dubois, Idaho. They hadn't isolated any organism and the blood tests were all negative. They didn't know the cause as yet, but felt it might be some new disease. The state and federal veterinarians were looking into some form of quarantine to prevent its spread.

It was a real puzzling situation to us all, until the old fellow who normally cared for the sheep came back after surgery and recuperation. He noticed that one of the vertical slabs in the fence was down, and the ewes were taking a short cut to the water trough by jumping through the gap. Being heavy with lamb, they were bumping the babies when they jumped over the pole. He got a hammer and nailed the slab back on the fence and the great abortion puzzle was solved.

A Few Drinks Too Many

"Dr. Dennis, my pigs are sick." That was all the information offered. I'd learned by now that a good history helped me find the answers, though some of the owners were reluctant to share everything with me. I could glean a lot of information on changes of feed, movement from one pen or pasture to another, how much time elapsed and other bits of information. Their own treatment of the animals was often withheld, I think because they were not sure if it was right.

"Have any of them died?" I asked. If there was a dead one, I could do a necropsy. I'd learned to rely a good deal on my own observations and findings on necropsies, because it was difficult to get laboratory help. By the time samples got to the lab two or three days later, they weren't much good.

"No, none of 'em are dead, but they're sure sick." I thought about it being hot. The weather often played a major role in the problems of animals.

"What are they acting like?" I needed some indication of the trouble. I think sometimes people felt like I must be stupid when I'd ask so many questions.

"Well they just want to lay under the shed. They squeal a lot when disturbed, and they won't come out and eat their grain." I wondered about erysipelas, which made them lame and off feed.

"Are they down and can't get up like they are paralyzed?"

"No. They get up when you poke'em, but they squeal a lot like they're hurting, and they stagger and fall down." I was weighing the lameness of erysipelas again.

"How many of them are sick?" Most diseases start with a few and get worse.

"All of 'em, the whole derned bunch." That pointed to a feed or water problem.

"Is there water in the pen for them? I guess they're in a

pen? I didn't ask you." It was summer and getting hot. I had seen pigs with heat prostration, but he said they were under a shed, and so probably in the shade.

"Yeah, they got water, and we feed them slop too." Still sounded like something they had eaten.

"Have you changed feed in the last day or two?" Sudden feed changes caused more problems than about anything else.

"Well, yeah, in a way we did, I guess. We've been gathering up slop and feeding them, but we haven't had much. A couple of days ago we emptied all of mom's old canned fruit. She had a cellar full of it and we needed bottles." At least I was learning a little.

"Did you give it all to them at once?"

"Oh no. We dumped it in our barrel and we been feeding out of that. There was a whole barrel full." I wondered if the barrel was sitting out in the sun.

"Where is the barrel? Is it out in the sun?"

"Yes, it's right by the pen on the south side." The light began to dawn. I could remember some drunk pigs when I was growing up. They were drinking the juice running out of a silage pit and the excessive squealing and staggering stuck in my mind.

"You say you dumped that fruit about three days ago?" In the hot sun it wouldn't take long to ferment.

"It was last Monday afternoon. This morning would be the fourth day." The time frame was about right.

"Does it smell like it was fermenting?" I needed to zero in.

"Yeah, it does have a twang to it, but I thought pigs could eat anything. They sure do like it." I was convinced, but the nagging possibility of botulism still was with me.

"One more question. Did you dump any green beans or vegetables, especially anything that seemed spoiled?"

"No, it was all fruit, no vegetables at all." That relieved my mind.

"Good, that relieves my mind. I'm pretty sure your pigs are plain drunk. That fruit has fermented there in the hot sun and made a lot of alcohol. Those pigs are soused from it."

"Well, I'll be! I never thought a pig could get drunk."

"How much is left in the barrel?" I wondered what they should do with it.

"Not very much, it's about all gone."

"Don't feed them any more for a day or two until they get sobered up. Then I think you can use it up. Don't feed too much at a time and you could dilute it with water." I thought it was safe enough to use.

"Maybe I should dump it out, there's only about five or ten gallons left?"

"Suit yourself, but I don't think you need to. I don't think there's anything in it to hurt 'em. That fruit with all that sugar is a perfect medium for the yeast to grow in, so it's alcohol you have to worry about. Just try to keep 'em sober. If they don't look better by tomorrow, give me a call."

The Ferocious Pig

It was late in the fall when the days were pleasant and the nights were getting a little nippy. I got a call from a Mr. Anderle who inquired if I castrated pigs. It was a simple straight forward request, but my mind flashed the questions. How come you're calling me? Doesn't everybody do their own?

"How many do you have?" I parried for more information.

"Just one. He's our herd boar, but it's time for him to go." He was big, I learned, and then I knew why he'd called for help. A lot of people were reluctant to castrate big ones.

"I see. Yes, I can come and do him. I can come this afternoon, if you want. Where do you live?" I thought I knew the place, but it's always good to be sure.

"Down the river east of Myton about six miles, then back toward the river, you'll see my place. I'll be looking for you this afternoon."

"I'll be over just after two. I've got to see another critter right after lunch." When it was time for me to leave, I gathered up my hog catcher and put it in the old Ford truck with the telephone body. I'd purchased the old truck from Crumbo Motor because of the body. It had 95,000 miles on it and the rear universal sang a loud song of complaint, but that body had a multitude of drawers, cupboards and shelves behind metal waterproof doors. The back was still open, so I could haul most anything, and it had a cover that could be pulled down to keep out the storm. It was perfect for a practice vehicle and I was thrilled to have it. When I put the hog catcher in, I knew I couldn't hold a big pig with it, but I felt I needed the tools of the trade. There was a scalpel in my surgical kit, but my hand instinctively touched my pocket knife. I always kept it honed and worked with a steel until it was as sharp as a razor. The scalpel might look a little more professional, however the grip on the knife handle plus that long pointed blade was my choice for this kind of job, where quick bold strokes reduced the pain, as well as the time.

As I drove southwest out of town, the sky was thinly clad in a high layer of haze, and I noted the sun had moved perceptively southward with the season. Out of my window a short distance from the sun was a bright spot in the sky with faint colors of the rainbow. It was a "Sun Dog," my dad had told me when I was a boy. He said the Indians called it a 'Sun Dog' because it was following the sun, and it always meant there was a storm coming. I thought that's probably right, because that high, thin layer contains moisture from an approaching front. It was the moisture that reflected the light from the sun. I thought about my dad and all the old timers and how they gleaned weather changes and other information from their own observations.

It put me in a reflective mood, and as I approached the Ioka turn my mind raced back to a day when we were coming the other way on this very same road in a Model T Ford. The road was newly made and covered with gravel then. We had the old Ford loaded with a tent, grub box, bedding and utensils for camping at the U.B.I.C. All the family was there and our spirits were running high with anticipation. Maybe a little too high, for that old Model T was clipping right along on that downhill stretch. All at once we heard a noise and saw the left hind tire still mounted on the rim had come off. It rolled past the front of the car and down the middle of the road. Dad stopped the car as soon as he could, but that tire kept rolling for a quarter of a mile down that road and finally went off into a gulch full of water. Dad had to wade out into the water to retrieve the tire, but all we had to do to fix it was pry up the back wheel with a post and adjust and tighten the clamps on that rim and we were off again. Our spirits had been subdued sufficiently that we went a little slower down that rough gravel road.

When I passed the site of the old flour mill north of Myton, I remembered the good feeling we got when we came there with a wagonload of hard winter wheat and traded it for flour for the winter ahead. There were the little sacks of germade cereal too, which were my responsibility when I was a small lad. I got to load them and hand them off the wagon when we got home. All the work of plowing, planting, watering, cutting and threshing somehow seemed worth it when we headed home with those gleaming white sacks in the wagon.

I guess I daydreamed all the way to the place, for when I pulled in, I realized I hadn't thought anything about how I was going to do that boar pig. I looked around for the pigpens and saw a sow and some little ones in a small fenced pasture. There was a small cow corral built out of heavy poles and huge cedar posts about fifteen inches in diameter. I thought, I'll bet those have been there thirty years and will probably last another fifty or more.

When I stepped out of the truck and looked inquiringly at the other pigs, the owner said, "He's in this pen," pointing to the cow corral. I retrieved the hog catcher from the back of the truck and approached the fence. It was a good solid pole fence about five and a half feet tall, and I was glad it was. When I stepped up and looked over, that old boar charged at me like a belching steam engine. He was huge and his staccato roar was about as loud as a lion in the circus. I estimated his weight at nearly six hundred pounds, and it wasn't hard to see that he was mean. His tusks stuck out over four inches on each side of his mouth and they were sharp.

"We can't do anything with him," the owner said. "We can't even use him to breed any more, because we can't get the sows back out of the pen, and we're afraid if he gets out, he might kill something." I remembered when I was in my teens how the Peatross Holstein bull had had his guts ripped out when he picked a fight with a big boar pig.

"What are you going to do, cut him and then sell him after he heals up?" I was wondering how we were ever going to catch and handle that mean bugger. I thought if he was going to sell him, I might encourage him to do it as he was.

"No, I thought I'd cut him and then in a couple of months when the smell leaves him, we could make him into sausage."

"You sure couldn't eat him now." That pungent male odor radiated from him clear out to my truck, and the more he got stirred up, the stronger he became. He would stand there with his mouth open and roar at us, then back up a little and charge the fence again. I looked at the hog holder in my hand and went back to the truck to get my ropes. I stepped up on the fence with a loop in my hand, and this taunted him even more. I was trying to figure out how to get that loop around his snout and flipped it at him a time or two. I don't know whether he was getting tired of telling me he was boss, or whether his bark was worse than his bite, but he calmed down some as I taunted him.

"Do you have any grain?" I asked, thinking if I could get him eating, I might slip the noose over his nose and back behind his tusks. I'd done that before at home.

"I've got some corn."

"That would be good. Why don't you get a little pan full while I keep trying to make friends here?" The old boy was definitely getting less defiant as I teased him with the rope. I moved up the fence to one of the posts and threaded my rope through about halfway up and wrapped it around once, then I took another wrap above the next pole. I knew we'd need at least two dallies to hold him when he threw all his weight around.

"Okay," I said when he returned. "Pour that corn close to

the fence about a foot or so past this post." I pulled through enough rope to make a loop that would reach the ground. When that corn was poured in, the old ferocious man-o-war just became a pig again and began to gobble it up. I dropped the bottom of the loop right in front of the big round snout and drug it back into his mouth in one sweeping pull. I think his tongue pulled it back past the tusks. I pulled up and tightened the loop. About the time he felt the pressure, I pulled the slack around the post and braced my foot. We had him caught. He reared back with all his might and threw his head from side to side, then squatted down as he hung back. The squeal that erupted wasn't too unlike the steam engine whistle I remembered; it hurt our ear drums.

I had him caught—now what was I going to do? On less monstrous models, I had the owner hold up on their tail while I pushed them against the fence. On this one, that was impossible. Maybe I could get a rope around his belly in front of his back legs and pull him up sufficiently to do the surgery. We had the one rope tied, so I made a loop with the other one, and dropped it down over his butt. It took some prodding and pulling to get it under his feet, and then we pulled it tight in front of his hips. By taking a wrap on the top pole and some more poking and pulling, we had him over against the fence, but he was sitting right down and we couldn't lift him. Now what? I spotted the big hook trailer hitch on the back of my truck, and the light came on.

I left the owner holding the rope, while I backed my truck up to the fence. We kept as much tension on the rope as we could while we undid the wrap on the top pole and tied it to the trailer hitch. With a little mechanical power from the truck and a good strong fence pole, we hoisted him up till his feet barely touched. It was a simple job to reach down over the fence and perform the required surgery. What at first had seemed almost impossible, turned out to be quite simple, thanks to a strong fence built by a pioneer.

Now, all I had to do was get the ropes off. By pulling the end through the loop it wasn't hard to get the back-end rope free. Then I used a trick I'd been shown many years before. There was a strand of bailing wire handy, and I threaded it through the loop on his snout and wrapped it around the post. Once we let the rope loose, he wasn't very long backing out of it. The constant squealing immediately stopped and we shook our heads, because our ears felt like they were full of water or something akin to it. By the time I got my hands washed and my ropes gathered up, the old pig was back eating the corn.

"Boy, he'll make a lot of sausage and lard," I said as I tried out my voice and hearing again.

"That's right. By the time he gets that smell out of him it will be winter, and we can keep the meat frozen. You know it takes two months or more for them to get so you can eat them."

"I know, we used to do the same thing when I was a kid growing up, but we never had one as big as that old monster."

"What do I owe you?"

"Let's see, it's about fifteen miles. How about seventeen fifty. Does that sound fair?"

"Sounds all right to me. I'll get you a check."

The sun was dipping toward the tops of the mountains as I drove back up the road to Myton. The cloud layer was a little thicker now, with wisps and ripples starting to form. "The Sun Dog" was long gone and I felt good about how things had gone. Taking one step at a time, using my head with a little persevering and ingenuity, I solved what first appeared to be quite a problem.

15
HORSES, COWS, AND PROBLEMS

Colic

"Doc, this is Les Mullins. My old stallion is sick. I wonder if you could come out?" Les had a furniture store in town and his hobby was breeding and racing horses. We had a lot of quarter horse breeders around the area, but Les was a thoroughbred man.

"What's he telling you, Les?" He lived less than a mile out of town, but I wanted to come prepared.

"I think he's got the colic, Doc. He's been laying down some, and he turns his head around to his side and stamps his foot."

"Kind of sounds like it. I'll be right out." Les lived just over the hill on the Neola road and he had a well-kept set of corrals and pens for his horses. They were rough-cut plank fences all painted white, sitting against the hill on the south side where the upper Cove Road comes in. He ran the store during the day and took excellent care of his horses before and after work.

"Why don't you pull right in here and get your truck off the highway." He was standing with his hand on the open gate. I could see the dejected looking stallion and pulled over toward his pen. The ground was a little moist from a rain the previous day. "He isn't as bad as some of them get, at least not yet. I noticed he didn't look right when I came home from the store, and I came right over. He keeps turning his head around to the side and pawing with his front foot. He's laid down a couple of times too." I could see the hair was a little wet on the side of his neck and behind the front leg, indicating some sweating, probably from pain.

"Have you exercised him at all since you got home?" I knew exercise was a good first step and many times fifteen minutes of jogging cured some colic. I always recommended it to owners that lived out of town. Then asked them to call me again if their horse still showed symptoms and hadn't passed anything.

"No, I just got here, but he usually doesn't need any; he runs himself to death around that pen. He gets himself so worked up he's in a lather, especially if I've got a mare in heat." He was in a good-sized pen. I walked over and checked the membranes in his mouth; they were still pink. I listened to his heart and moved my stethoscope back along his abdomen checking for gut sounds. There wasn't much going on in that food factory that I could hear. He did turn his head as I heard a muffled groan in the bowel.

"Let's move him around here a little, Les, and see if we can stimulate his gut. You lead him and I'll pound him on the tail if we need to. It's better if he jogs, if you and I can stand it."

"Come on, old boy. I'll take him down toward those mares. I usually have trouble holding him back." He hooked the lead rope in his halter and started off. The old guy followed on a walk, but he didn't want to trot very much. With me tapping him with a piece of wood, we did get him to jogging now and then. After about five minutes, he was hurting a lot more and wanted to lay down.

"I think we better get some oil into him, and I'll give him a shot for pain." I was convinced he needed to be treated. He laid down and threatened to roll, and I was wishing I had treated him before. I went to my truck and poured a half gallon of mineral oil in my bucket and added an ounce or two of turcapsol. Then I drew up the pain shot, picked up my stomach pump and tube and went back to where he was laying. We urged him back to his feet, and he didn't even pay any attention to the needle I popped into his jugular vein. I gave him the pain medicine and Les held his halter as I started the stomach tube up his nose. He didn't like that and pulled back, but I was able to keep up and slide the tube back to his throat. I had to try several times to get him to swallow the end of the tube. If you merely push the tube in, it slides over the epiglottis and into the trachea, or windpipe, evoking a cough most of the time. By wiggling the end of the tube back and forth, it started him to swallow, and I thrust it forward into his esophagus. It felt like there was more resistance and as I pushed the tube I could see it passing down the side of the neck. It had to be in the esophagus if I could see it, so I knew I was in the right place. It was the key I always used when tubing horses, and prevented me from pumping some foreign material into the lungs. This could be a fatal mistake.

When I had the tube in the stomach, I had Les hook up and pump in the oil and medicine. We then walked the stallion for about twenty minutes. The gut sounds increased and the heart rate speeded up some as well, but no gas or fecal material was passed. I got a sleeve from my truck and did a rectal exam. The bowel was empty as far as I could reach, and the colon flexure didn't feel impacted. There wasn't much more we could do except we did give him a small enema. It was getting dark as I left for home, reflecting on the situation. It was my conclusion that the storm had softened some otherwise rejected hay, and he had tanked up a little on it. I thought he must be plugged off some place, and hoped that it would work loose.

The next morning I went back out to see the stallion, and a funny sight was awaiting me. There was the old horse with his butt on the ground and his front legs holding up his front end like a dog sitting down. I kicked him up and checked him over. He was about the same as the night before. When I finished giving him another shot in the jugular vein, I made him walk for awhile, but he soon wanted to lay down. I could see by the bruising of the head and around the eyes, that he must have been down during the night. I was worried that he was going to die. When I started to leave, he rolled up and pushed his front feet out straight like he was going to get up, then just sat there. I had never seen a horse do that, and haven't seen another since. He must have received some relief in that position and maintained it for the next five days. Each day I expected to find him dead, but there he would sit like a dog. On the afternoon of the third day, there was a small amount of oil seeping from the anus. It gave me a ray of hope to go along with three days of survival.

"Has he eaten or drank any water?" I asked Les as we stood watching.

"Not one bite that I know of. He won't touch any grain. I did see him smell the water, but I don't think he drank any. What do you think, Doc?"

"I suppose it's anybody's guess, Les. I didn't think he would live this long without passing something; however, I'm encouraged by the oil coming through. Have you heard any gas or seen anything come?" I wondered if he might have seen something I wasn't aware of.

"No, sir, and I've watched all over the place. Have you ever seen one sit like that before?" He shook his head slightly in disbelief.

"I never have, but he seems to get some relief from being in that position. At first I thought he was just too weak to get on up, but it must feel better that way. I guess we'll just have to wait and see what happens." He sat like a dog for two more days, then began to stay on his feet and sip some water. It took him a good two weeks more to get back to eating normal again and he fully recovered.

I suppose the severity of symptoms and the low rate of success in the treatment of colic in horses had an influence in my being called on for help rather early in my career here in the basin. No one likes to see any animal suffer, and a horse with a bellyache soon evokes concern. Much like the newborn babe, the horse complains with characteristic symptoms. He doesn't cry, but tells his story by sweating, turning his head back to his side, stamping his feet, and finally flopping down on the ground and rolling. Horses appear to have low pain thresholds with inflamed swollen limbs and especially with digestive problems. When those griping pains set in, they are in real distress and many of them try to roll up on their backs, which apparently gives them some relief.

There are many causes for colic in the horse, and for me it was sometimes difficult to put my finger on what the problem really was. It could be a plugged location in the gut, a twist or torsion often involving fat globs in the omentum, thrombo-embolic colic where a portion of the gut dies, or simply diarrhea. Worms are thought to play a major role, because larval stages of some of them invade and live in the arteries supplying all the intestines. If a blood clot caused by the larvae plugs a vessel, that portion of the gut dies. With the anatomical knowledge that I had, I could sometimes get an idea by listening with my stethoscope on various areas of the abdomen. I always did a rectal exam, and occasionally found an answer that way. If it was simply a dry hard plug of the terminal bowel, I could remove it, and a few times I was able to massage and break up a blockage further up the gut. The sickening feeling of fecal material outside the bowel indicated a rupture, usually of the stomach, and I prevented further suffering by putting the animal away. Prognosis was primarily based on heart rate increases and the color of the membranes.

I didn't have a bag of magic tricks that owners wished I had, however my knowledge exceeded their own, and they depended on me. For treatment, I stuck to the basics of relieving pain and relaxing the gut with the medicines I had and pumping some mineral oil with gas suppressants into the stomach. Most of all, I felt jogging to stimulate natural bowel activity was paramount. The low ratio of success, combined with the high cost of abdominal surgery deterred me from entering that arena, but I often wished I could have seen inside. I certainly didn't save them all, and I suffered much anguish as I watched them die, but I knew I'd done the best I could.

Rain Dance Deer Horn

The pickup truck pulled off the highway and parked out by the sign posts. It was a hot day in late summer, and the effects of the draught were evident on the faces, as well as the pastures. The hot winds had parched the grass until the pastures were brown, except along the head ditches where scalloped patches of green showed where the last water turn had crept down a few rods before drying up.

Benton Ridley got out of the truck and came in through the door. He was an Indian friend I'd known for a decade or so. Over the years I had developed some cherished friendships among the Indian people, and he was one of them. They were always quiet and attentive as they listened to my advice and counsel. I wasn't sure of how well they followed what I told them, because they seldom shared what happened at home with me, but they kept coming back and we were friends.

We had the swamp cooler on and the room was much more comfortable than the outside. Benton always wore a tiny smile and he held out his hands toward the air coming from the cooler, much like someone would warm their hand at a stove.

"Hello, Ben. What do you need today?" I said after getting up from my chair in front of the desk. He took his time coming up with his answer.

"Maybe I could get a job. Sit here all day where it's cool?" It was spoken in an assumed seriousness only betrayed by the corners of his mouth. "How much you pay?"

I grinned a happy grin.

"Not a bad idea, Ben. I'll pay you the same as I get for sitting here with nothing to do." I welcomed Benton's needling, but felt a need to let him know our plight. He nodded his head up and down to let me know he understood.

"I got cow with bad foot. Swelled up. Limps." He'd said it all in those few words. It was so typical of all the Indian clients.

"Probably got foot rot. You need some pills." I tended to follow his lead with few words. Again he gave his nodding agreement, and I went into the drug room to get the pills. Knowing what kind of cattle Benton had, I got the dosage I thought he'd need and returned to the counter. "Big cow?" I asked and he answered with the nodding. "Give her all of these at the same time." The phone rang. "You might want—" The ringing continued. "Excuse me." He nodded again.

"Basin Vet Clinic."

"Doc, this is Ray Brown. Could you come up and look at a cow for me?"

"Yeah, sure. What's her trouble?" Benton was listening. Everybody seemed interested in other people's livestock troubles. I guess they wanted to learn, and this way it was at someone else's expense, not their own.

"Well, Doc, it's the damnedest thing I've ever seen. She's swelled up in the face and it don't look like she's ate or drank anything for a week." Swelled up in the face. My mind was racing. Ray would know lump jaw. It could be a rattlesnake bite. I'd seen pictures of horses. It had to be some form of lump jaw infection in this area.

"Is she swelled any under her jaw, Ray?" I wanted a better clue.

"No sir, Doc. It's her mouth, her upper lips. They're all pooched up and swelled. I've been in the cattle business all my life, and I ain't never seen anything like it. I think you better come up." It really sounded more like a snake bite, but I'd never heard of any, and we seldom ever see a snake anymore. I was both puzzled and intrigued.

"I'll come right up, Ray. Where is she?" I knew he could have cows in several places.

"I've got her in Vic's big corral. You know, in that field east of his house. I'm here at home in Neola now, but I'll meet you over there."

"Good enough, I'll see you in a little bit." I turned back to Benton and the curiosity showed in his eyes.

"Ray Brown. He's got a cow all swelled up in the face. I got to go up and see her." I shared that much of the other side of the phone conversation. "I started to say, maybe you ought to take a good look at that foot. It could be a thorn or a nail." This time he shook his head the other way.

"Already throwed her down. Couldn't see anything." That figured, Benton was a good cowboy and the old philosophy that, if it couldn't be done on the end of a long rope, it wasn't worth doing, showed forth in his action. This time I nodded my head in approval and he picked up the little box of pills and turned toward the door. Then he stopped and turned back. "How much?"

"Ten dollars and fifty cents," I said as pleasantly as you can tell people how much things cost. He reached in his pocket and pulled out a few greenbacks folded together and slipped a ten out of the ones. Then he went deep into his pocket again and came up with some change. Picking out two quarters he laid them and the ten on the counter. He looked outside again, then hesitated as Nancy was writing him a receipt.

"Hot out there." He pointed with his box of pills.

"Yes, and plenty dry too. Thanks, Ben." I handed him the receipt and a funny thought flashed into my mind. "You know what you ought to do, Ben, is get a bunch of the Indians together and have a rain dance. We could sure use some rain." I said it jokingly to him. I've never talked much about their traditions and sacred dances, like the sun dance, for fear of offending them, but I'd never heard of the local Indians having any rain dance. It was the Indian lore we read about in books, and I was pretty sure it wouldn't offend Benton. He didn't say anything for a moment or two and then a smile broadened his face and he said, "White man's turn." And walked out the door. Nancy and I had a good chuckle at his pointed humor.

When I crested the big hill, I could see Ray's outfit already over by the big corral, and when I pulled in, he and a neighbor were on the corral fence looking at the cow. As I joined them, my first impression was how gaunted that poor old cow was. She was so empty her sides were almost touching. I climbed the fence and walked out toward her in the big corral. She staggered as she turned to face me and acted a little like she wanted to be on the fight. There was the swelling Ray had described, but it was more localized than I'd expected. I thought the whole top of her nose was going to be puffed up. It was just the corners of her upper lips. Ray was certainly right when he said it was strange.

"Well let's run her in the squeeze chute and see if we can figure out what her trouble is." I knew I'd need to look closer than what I could see at a distance. She stood her ground for a second, but with three of us waving and shouting at her she turned and about fell down.

"I hope she can make it. She seems a lot weaker than when I brought her in," Ray remarked as we followed her through the gate. I thought the exertion of walking to the corral had probably used up all of her reserve, but she got better as she went.

"I'll get my nose tongs while you push her in there," I said as I made for my truck over the fence. When I grabbed her by the lip to put the nose tongs in her nose, I felt something hard up under there. I pulled her head up a little and tied the rope. Then I slid my finger under the side of her mouth and took a look. There was a deer horn. I carefully pushed the lips out over the points on both sides and pulled out the horn. It was one side of a large two point buck and the shaft of the horn had been down the cows throat, with the two points being caught under and pushing up the lips.

"That ought to make the old gal feel a little better," I said as I held up the horn.

"Well, I'll be damned!" Ray exclaimed in astonishment. "Who'd ever thought a cow would get something like that caught in her throat? That thing's eighteen inches long." He kept shaking his head in disbelief.

"Boy, it's amazing what a cow will eat when they're mineral hungry," I offered, thinking of all the things I'd found in their stomachs.

"That's right. It's my own fault. I haven't had any mineral out for those cattle all summer. I'll have to get some." Ray was chastising himself.

Feedlot Founder

Early in my experience in the basin, I got a call from Zane Christensen up in Talmage. He said he had some steers dead and some that were down. He said they were in the feedlot on barley and hay. The red flag in my mind went up, because I knew that a combination of barley and alfalfa often led to bloat, especially when you've got above four pounds of barley. I asked him if he thought they were bloating and what his ration was. He told me they were on a fairly high level of barley, something like eight to ten pounds and alfalfa and grass hay but he was sure it wasn't bloat because they had been on the same feed for more than a month. He said there was some that were loose in their stools and quite a few down. He wanted me to come up and see if I could tell what the trouble was. I'd never had any experience with a feedlot situation, but I knew from school that it could be very serious if some disease began to spread in animals that were closely confined. I couldn't think of anything special to take, since I carried quite a lot of drugs in my outfit. I'd asked him if the ration had been changed in any way, and he said it was just the same. The only other thing that came to mind was a poisoning of some sort.

It was late in the afternoon when I got there. I looked things over and opened up one of the dead steers and it all pointed to a grain overload. I told him this, and once again he reiterated that nothing had been changed, that it was exactly the same grain ration.

I was convinced, so I began to treat those that were down with calcium gluconate mixed with physiological saline and dextrose solution in the vein. They responded by an increase in intestinal and stomach motility and became more alert. Some even got up and most had watery loose bowel movements. It was time consuming, because the calcium had to be given slowly and we worked on into the night. I had twelve bottles of calcium and some other fluids and one by one we treated the worst cases. I ran out of medicine sometime after midnight and there were still steers down. I offered to go to Roosevelt for more medicine and return, but Zane said, "Let's wait until morning and I'll call you if I want you to come."

As we worked that night, I learned that he had gone over to a place near Craig, Colorado, and brought back a big truck load of freshly harvested barley. We came to the conclusion that somehow it was responsible. The amount hadn't been changed but the quality was better and it was freshly harvested.

The next morning Zane called and said things looked much better, most of those that were down were up and seemed to be improving. It was a great lesson for me, and I've recommended hundreds of times since, that when you change feed, don't do it all at once. Mix the two feeds together to start with, and make the change gradually. Most of the time, it has been after someone had had an adverse reaction, but hopefully it has helped them in the future.

I've done a lot of work for Zane down through the years and held him as a great friend. I've felt he has thought well of me and my service, because he seemed to trust me. One time he took me over to Ault, Colorado, to pregnancy test some cows he was buying, because he wanted to be sure what he was getting.

Equine Injury

Challenges in my practice came along often enough to keep it interesting and one of these was a call from Milt Larsen in Arcadia. He said he had a colt that had cut its head real bad and he didn't know whether to kill it or not. He wanted me to come up and look at it. When I arrived, there stood the horse with about half of the skin on its head hanging down around its nose. Nearby was a Minneapolis moline baler that he had caught his head in. The feeder mechanism of those balers had some long sharp teeth pointing downward that pushed the hay up into the mouth of the baler chamber. They were about nine or ten inches apart each way. The two-year-old colt had stuck its head down between the teeth to eat some hay that was in the feeder mechanism. Whether it got startled or something, Milt didn't know, but two feeder teeth caught the skin just above each eye on the front and one or more had caught the skin at the angle of the jaw at the throat latch level. When he jerked his head out, it ripped and stripped all the skin off his head from the eye level down to near the nose. There the horse stood with that shredded skin hanging down over his mouth and nose.

My first thought was that it probably wouldn't heal too well if I did sew it back, but poor old Milt was feeling so bad I couldn't bring myself to the point of recommending it be put out of its misery. It seemed to be a special horse to him. There were the remains of an oat straw stack close by, so I suggested we lay him down on it and I'd see what I could do to patch him up. We took the top off the straw and it was fairly clean, so we led him up to it and I gave him a general anesthetic. I cleaned the skin and stitched it back in place. I'll bet there were 150 to 200 stitches over the pony's face and head, but all the skin was there and it all fit back together. By comparison, the horse looked great and he got back on his feet before I left. I didn't get a chance to check on him much during the healing process, but about a year later, I saw him and there were just a couple of hairline scars where some of the skin died and sloughed off. It once again reaffirmed my natural tendency to not give up and try to do something even when it looked hopeless.

Another, what could be termed, horrible case, was a horse that belonged to one of the Indian cowboys. While chasing a cow his front feet had come down on the end of a stick; the other end flipped up and caught the horse just behind the penis. It had gone clear on through between the legs, coming out very close to the anal opening, then ripping out everything in-between. The horse's hind legs were literally split in two, but the big vessels had escaped somehow. It happened up in Coyote Basin and by the time I got there, the old horse was in shock and trembling. The wound seemed fairly clean and could be left to drain, so I just deadened the skin edges and began to sew it up from the top down. As I got down between the legs, we sort of pulled his legs apart so I had room to work, and I knelt down under him and kept on stitching. I left a small opening where it drained for a few days and he healed up as good as new.

Sometimes things looked a lot worse than they really were, especially to the owners, and this was the case with a horse brought into the clinic by a man and his family. It had run past something sharp that cut it across the ribs and side in a big half-moon-shaped cut about eighteen inches long. It was swelled and pulled apart about six inches and crawling with a few hundred maggots. They wondered if I could sew it up. After that many days and with all the swelling, I knew it couldn't be expected to heal back together by suturing. I cleaned out the maggots and fly eggs and told them it would take some time, but it would heal.

"But, what about the big scar?" he said.

I assured him that it would heal with only a small scar, but he was still very dubious. I sent some medicine home with them to spray on it every day and told them to bring it back once a week until it was healed. They were amazed at what happened, and after the second trip didn't bother to come back any more. They later told me you could hardly see any scar.

Another amazing case was a colt Erv Murphy brought into the clinic. It was about nine months old and still with its mother. It had a tree limb about an inch and a half in diameter stuck up into its abdomen and up under the ribs. The rest of the limb, about six to eight feet long, with all its branches, was sticking back along the side of the colt. We led the mare out of the truck and down the unloading chute, with the colt following along. The tree limb didn't even hardly hit the sides. We were elated, because we thought if that limb gets pulled out, his intestines will come out through the hole, and we would be in a mess. I got everything ready to operate and sew him up out there by the chute. We planned to push him up into the chute by his mother. He was a wild little rascal and had never been caught in his life. When we were all ready, I got in the pen behind him, and Erv led the old mare about a length or two down the chute. When the colts head went past the edge of the chute, I sprang forward and grabbed the branches of the tree limb and pulled it out of his tummy. To my surprise, nothing but a little fluid came out of the hole. We crowded him up behind his mother and got a pole behind him. I gave him a tranquilizer and when he was pretty well out of it, I examined the wound. I found the limb had slid along the inner side of the abdominal wall and up under the ribs, without any apparent intrusion into the cavity. I left the hole open to drain and gave him a little antibiotics. He went back to the field with his mother, without the tree limb sticking out behind.

Imbedded Foreign Bodies

Imbedded foreign bodies, mostly of the wood variety, were strange and different in their outcome. I remember some horses with relatively small wood slivers in the upper muscles of the hind legs. They became infected and died in a couple of days.

Then there was B. T. Frost's horse. I was called to come and castrate the horse and told when I got there that six months or more before he had stepped on a stick that flipped up and punched a hole just lateral to the testicles. It didn't seem to bother the horse, but the hole had never completely healed and had a little drainage from it. I gave the horse some anesthetic and we tied up the legs because I wasn't too eager to open up a castration wound where there was infection. I wanted to explore the wound that didn't heal, thinking there might be a small sliver up in there. I began to explore in the opening and found a tube going straight up toward the pelvis. By cutting and enlarging the tube, I felt something hard with my forceps and I finally got my fingers in to touch the thing. It was a round chunk of wood about an inch in diameter, but it seemed quite immoveable and attached. I got some towel clamps hooked in it and pulled quite hard and it began to slip. When it came loose, there were two pieces, each about five or six inches long, hooked together by a green break. It had gone up into the horse, hitting the bottom of the pelvis bone, where the one piece broke over and went back along the bottom of the pelvis. I could hardly believe my eyes, when I saw that big chunk of wood which had been in the horse for months and months and hadn't killed him.

I had a similar case with a cow with a wound draining on the side of her neck. When I got the piece of wood out, it was about ten inches long and the size of a shovel handle.

Fatal Slip

Sometimes the impossible happened, and the animal would walk away or recover, and other times, it would seem to be an impossible little accident, and it would cost their life. I was called to go see a horse that was down and couldn't get up, on the upper road to Pleasant Valley. Paul Hansen was taking the horse out to his sheep herd and leading it behind his pickup. It was trotting as he slowly drove along the dirt road, and as he went down a little slope, the horse seemed to slip and fall and couldn't get up. When I got there, the horse showed a definite posterior paralysis, but I couldn't tell why at first. The hind legs and feet wouldn't respond to any stimulus. I wondered if an aneurism had come loose and plugged the bifurcation of the aorta, but he should have been able to feel some pain. By carefully feeling along the backbone, I felt a definite ridge just in front of the pelvis. He had broken his back, somehow, by slipping and falling.

Saddle Sores

This was a condition related to sinus of the withers because of its location, only it was on the backs of the tall spines of the vertebrae, and not the front. Saddle sores were seen most often on horses that had been ridden on pack trips into the high Uintahs by the inexperienced or periodic riders. It was the necrosis or death of tissue over the tops of the spines of the withers caused by prolonged pressure from inadequate padding of the saddle. The insufficient padding of the saddle was exacerbated by the weight of the rider being thrown forward as the horse went downhill. When the front of the saddle rested on the bony prominence of the withers, the skin and tissue underneath were deprived of a sufficient blood supply, and they would die. By the time it was noticed or any swelling occurred, it was already very painful to the animal. A treatment by the old timers of putting wet salt packs on the area in the very early stages helped to reduce the swelling and soreness, and if the saddle was padded so it didn't ride there the horse could still be used. In the more advanced cases, the necrosis had already occurred and the only treatment was to remove the dead tissue and dried-up piece of skin and let it heal from the inside out as an open wound. This was painful to the horse and they didn't always take it without letting you know. After the wound was open, we used some kind of medicine like scarlet oil to stimulate granulation and healing. It would take weeks for them to heal, and months for the tenderness to go away.

Pipe Ring Bracelets

Odd things sometimes happened to animals, and pipe ring bracelets was one of them. When oil was discovered in this area, there were many miles of pipe used to pump water to the wells for drilling and for carrying the crude oil to the tank batteries. To protect the threads on the ends of the pipe during transportation and handling, a steel sleeve was screwed onto the pipe. Naturally, when the pipe was used, they became a by-product and were abandoned, carried home by workers, and sometimes bounced off along country roads. At any rate, some of them found their way onto the feet of some of the younger animals when they would step down into one of them. The weight of the animal would push the foot down through them until it wedged and wouldn't come off. This would cause swelling in a very short time and a painful lameness. To get them removed was a challenge, because simply trying to pull them off would damage the feet and pull off the hoof.

When they were brought in to me, I would put them on the operating table and tie their feet as solid as possible so I could work on them. Obviously, I couldn't use a cutting torch because it would burn the animal, so I had to cut the rings in two with a hacksaw, or emery wheel. This was a very tedious and time consuming task. Sometimes when cutting with the hacksaw the metal would get so hot it would make the animal squirm, and the emery wheel was much worse. I would have to cut a little, then pour water over the area, and go at it again. Since they were rather heavy metal and curved, it nearly always required another cut on the opposite side to remove them. Removing one of those steel bracelets could consume a big share of an afternoon.

Wounded Cows

Hey diddle diddle, the cat and the fiddle, the cows tried jumping the moon. I saw many wounds and cut-up cows as a result of jumping or climbing through fences, or crashing into wood piles, machinery, and many other obstacles. The thicker, tougher hide of the cow was not nearly as vulnerable as the horse's, but occasionally it gave way to create serious wounds. Sometimes bones were broken necessitating slaughter, but many of the wounds were fixable.

The most common request for my help came when a cow would tear up her teats and udder, spilling much blood and milk. Not infrequently, the large so-called milk veins in front of the udder would get cut. This made the owner frantic, because he would have to hold pressure to control the bleeding while someone else called me. In the early part of my career, I received a good suggestion from a colleague at a state veterinary meeting. He said when he got calls on animals bleeding, he would tell the owners to find an old spring-type clothespin and pinch the wound edges together and put it on until he could get there. It was surprising how well it worked, and I used the same advice.

The worst thing was the home remedies that were used to try and stop the hemorrhage. Most of the time their effectiveness was minimal, and all you ended up with was a well-contaminated wound. It seemed that flour was the most commonly used remedy. Most wounds were just skin and muscle, but occassionally they were deep enough to enter the abdominal cavity

The Oman family from Altamont called in to see if they could bring in a cow that was badly cut. When they arrived the patient was an old jersey milk cow with a fifteen-inch gash along the side of her lower abdomen. It looked like some of the stomach or intestines were showing from the back of the wound.

"What happened to her?" I asked as I straightened up from the examination.

"Well, you see, we had our old truck parked out by the corral and its front bumper is bent out a little on the end. We were trying to drive her through the corral gate, when she took a notion she didn't want to go. She broke and ran around the front of the truck cutting in too close when she made the corner. She let out a beller when she hung up on that bumper and then just stood there bleeding. That's when we called you."

"Looks like it cut through to the inside. I hope she hasn't torn a hole in the stomach or a gut."

"I never saw anything running out except blood."

"Well that's good. We'll have a look and stitch her up." When I washed her up and checked it out, there was just a little bulge of the stomach protruding and no evidence of any deeper injury. I was able to pull her back together with a few yards of catgut and stainless steel suture, but ended up with a tired back.

"There, that ought to hold her together until she can heal.

We got her corset all patched up."

"Yeah. You can hardly tell she was ever cut, except for the stitches. Thanks a bunch."

Heroic Effort, Sad End

Her voice was excited and halting, and I could tell she was almost in tears. "Doctor Dennis, this is Julie Farley. Our milk cow just jumped the fence and came down on a steel post. It jabbed up through the bottom of her and some of her insides are hanging out. Can I bring her over, or would it be better if you came over here?" In my mind, I could see intestines slithering out and the cow stepping on them, and at the same time, I could see myself trying to tie her down in the manure with not much more than a barbwire fence to tie to. How could I ever get things cleaned up, and back in, with that?

"Julie, what do you see hanging out?"

"It's a round ball about the size of a football or a balloon."

"Does it have any holes in it?" It sounded like there was some of the stomach protruding, and that would effectively plug the hole.

"No, I didn't see any. It's just like a balloon sticking down there. Do you think you can save her?" It was a thirty-mile ride but it would be a lot easier and cleaner here in the clinic. We could roll her up on the operating table, and it would be inside.

"How is the cow acting now? Does she seem weak or wobbly?"

"Oh, she's strong enough. She could run around when we were trying to get her back in the corral. She's just standing there now but she can move alright."

"Well, why don't you bring her over. I think we could put her on the table here, and do a better job of cleaning and sewing her up."

"Yes, I think so too. We'll be right over as soon as we can."

"Be careful loading her. Don't tear that stomach."

"We will. I think she'll walk right up the loading chute into the truck."

When they arrived, it was just as she had described it. A balloon of her fourth stomach was sticking right out of the bottom of her. If we could get her on the table, it shouldn't be too tough a job to fix her up. The moments were anxious, as we led her down out of the truck, but nothing touched the balloon, and we all heaved a sigh of relief. All we had to do now was lead her into the clinic and roll her up on the table. It was a simple procedure, but she had other ideas about that. The smell must have sounded a warning, or the cement floor frightened her. She planted her feet and refused to move. We had to drag and push her every inch of the way to get her standing beside the table. We put the ropes around her to tip her over and began to pull. Wouldn't you know it, she began to fight, and put her front foot up on the table and refused to be tipped over. When she twisted herself upright the balloon was torn half in two, and stomach contents and blood poured out on the floor. We got her back on her feet and clamped off most of the bleeders, then tied the whole stump with some umbilical tape. We wanted something to keep it from falling back inside of her once upside down. By tying her front feet with a combined effort we did roll her over and secured her legs.

All we had to do now was clean up the stomach, trim up the ragged edges and suture it closed. It was a long and tedious process to clean and turn the outer edges in together and suture it tight enough that it wouldn't leak. When we finished, we felt good about the results. We thought all we had to do was tuck it in and patch up the hole in the bottom of the abdomen. When we tucked it back inside, we discovered another small hole a few inches from the original. We managed to suture it and began cleaning out the contamination inside the cavity. As we washed and cleaned a little deeper, we discovered that the steel post had torn a lot of the intestines. All our hopes were shattered, all our effort for naught.

"Julie, she's all torn up in the intestines and the only thing to do is to have her slaughtered."

"Oh, no. Not after all that work." That's the way I felt. All that work for nothing.

"I guess you'll just have to trade milk for hamburger. At least it isn't a total loss."

Rectal Prolapse

I employed my own ingenuity sometimes to solve some problems. Whenever I had a rectal prolapse, I would turn it back inside and use a purse string type suture of the anus to keep it there. I'd instruct the owners to keep watch and see that manure was coming through. On many of these cases they would get plugged off, and then the animal would strain hard trying to push out the manure. All that happened was the fluid was squeezed out and the plugged stool got dryer and more packed. Many were brought back in when the owner couldn't get them unplugged. This was especially true with calves, and I was quite dissatisfied with the approach. I decided to amputate the everted portion and suture the ends together, especially on those that were torn. This was difficult to accomplish and resulted in a lot of infection, some of which was fatal. One company had produced some plastic rings with a deep groove in them for use on little pigs with similar problems. I took some plastic ABS pipe and filed a groove in it, but couldn't keep it in when the animal strained. I finally thought that if I could have two holes close together on one side, I might be able to insert a needle through from the outside and pull the suture material through the holes. Then, when I wrapped it around the outside and tied things off, it couldn't come out. For suture material I used small rubber tubing which I could wrap very tight and shut off the blood supply. By the time the tube and all sloughed off, in about five or six days, the serosal sides of the bowel were grown together, and the problem was solved. It has worked very well now for many years and I've passed the idea on to other veterinarians. The old adage of necessity being the mother of invention was true, and availability of many sizes of pipe for different sized animals fit right in.

Stolen Heifers

For several years, Parley Lambert had bought young Holstein heifers and raised them for sale as springers. He had some good pastures and with his prudent feeding, he had done very well. He called me late one morning. "Dan, could you come up and look at some heifers for me? Maybe I should say come and look at some skeletons and tell me what to do." That was all he told me and I thought perhaps he had some that weren't doing so well, maybe fluke or worm problems. Parley could be a little dramatic. When I arrived at his house, he came out, got in my truck and said to go south down the lane to his pasture. There were nine of the skinniest cattle that I had ever seen. Only a few were even on their feet. Most of them were laying down in the grass. The odd thing was they were reaching and eating all the grass within reach of their mouth. They didn't appear particularly sick, but they were very weak. I was puzzled and Parley could tell it, but he was one that could laugh at himself. He had a half smile on his face.

"What happened, Parley?" And this is the story that he told me.

"One morning as I was going by, it didn't look like all the heifers were in this pasture. I stopped and counted them and sure enough there were nine head missing. My first thought was that they had broken down the fence someplace and got out. I went all the way around the fence and couldn't find any place where they could have got out. My next thought was that the ditch rider or someone had left the gate open for awhile and they had got out, and I'd better go look for them. I drove up and down every damn lane in this part of the country and didn't see hide nor hair of them. I wondered if any of the neighbors had seen them, so I asked all of them I could find. They just seemed to have disappeared, and I didn't sleep much that night. The next morning I went back to check on all the gates, and on that south side by the road, I thought I could see some tire tracks. They weren't very fresh, but I didn't know how long it had been since they had disappeared. Apparently somebody had pulled in there and loaded those heifers. I called the sheriff and he was convinced that was the only way they could have gone. He alerted the brand inspector and they passed the word all over the state, and notified all the livestock markets. The highway patrol was notified. The next week didn't bring any reports of anything even suspicious, so we figured they must have been taken to California. The word got around to watch out for some slick rustlers. It had been over two weeks, and I'd given up all hope. Then this morning I had a water turn. I started it on this pasture and walked over by the old house. I noticed the door was shut and I'd propped it open so the cattle could get in there and shade up. Well, I found my heifers and damned if they weren't all alive. When I opened that door, they could barely get to their feet. They could smell that water and they went along sucking it up out of the grass like somebody drinking hot coffee. I thought if they got to the ditch and tanked up, it might kill them, so I hurried and turned off the water. That's when I went and called you. I'll be damned if I hadn't rustled my own heifers."

"It looks to me like they might have sucked up what water they wanted. At least they're eating and that's a good sign." I told him the pasture had been eaten down by the other heifers he'd moved, so I thought the best thing would be to leave them alone. They couldn't overeat in there, and they were so weak they wanted to lay down and eat anyway. "Why don't you turn the water back on that grass, and keep an eye on them. I think they'll be all right if they don't get down in a ditch and can't get out." We checked the old house and they had eaten every spear of manure they could get to and even chewed on the wood. I guess the coolness of the rock and brick had helped them survive.

Hard Lump Choke

A truck pulled around to the back of the clinic and I walked back that way. The driver backed up to the unloading chute and proceeded to unload a cow. I hurried in the corral and shut the big gate by the shed to keep her in the catch pen. Then I waited to have a look as she walked down the ramp and past me. He was right behind her and after closing the little gate turned to me.

"Doc, she won't eat and she's real sick." I think what constituted real sick was how gaunted up she was. I'd noticed she was plenty empty as she went by.

"Run her in the chute and I'll catch her head and check her out." I went down to open the head catch on the squeeze chute. He got behind her and she walked right down and put her head through. I closed the bars on her neck and noticed a little green liquid run out of her nose as she fought the confinement. The mucus dried around her nose was stained green as well. I'd had some experience with sugar beet tops getting stuck and knew it was a telltale sign of obstruction, or choke as we called it. In animals, that meant obstruction of food passing through the esophagus to the stomach. When she was settled down, I could see a swelling down on her neck at the entrance to the chest cavity. He had joined me at the front of her.

"What's she got, Doc?" He asked like I had a crystal ball I could look in and see all the answers. At least I'd seen enough that I had a good idea.

"I think she's got a choke, something caught in her esophagus. There's a swelling there in her neck." I pointed to it and reached down to try feeling it, but the cow didn't want me doing that. I caught her with my nose tongs, pulled her head up and around and wrapped the rope to hold her. I felt the lump. It was hard and it wouldn't move. It didn't seem to have much feed packed against it. When I squeezed it in my hand, trying to move it, the old cow told me it hurt by dancing and fighting. I thought back to the beet tops we used to push on down into the stomach with the Kingman tube, a big stiff stomach tube. "There's something caught in there and it feels pretty hard."

"Can you get something like that out?" He was feeling the lump.

"I don't know for sure. We used to push beet tops on down

into the stomach when I was in Idaho. I think we should try that on her." I'd already decided to try that.

"Sounds good to me." He knew what was wrong now, and hoped it could be fixed. I went into the clinic after the Kingman tube and returned.

"I'm going to release her head and I want you to hold the rope out in front of her while I put this tube down. We need to have her neck as straight as possible." When he had the rope held across his hip, I put my left hand over her nose and in her mouth on the far side. This way I could help hold her neck straight by pressing her muzzle against my hip. I pushed the end of the tube over her tongue and down her throat with a minimum of fuss until it hit the obstruction. When I got a new grip on the hose and put some pressure on, nothing moved and the old cow really began to dance with her front legs and to fight. I had assumed it would have to hurt some, but the thing to do was have her pay the price and force it down into her stomach. I decided to put on all the pressure I thought the esophagus could stand without rupturing, and let the cow suffer. I held on to her mouth as tightly as I could and quickly pushed as hard as I dared. It didn't budge. The poor old cow really told us it was painful, so I pulled the tube out. There was a little blood on the end of it. I was stumped now, as I stood looking at the situation. It was never advisable to cut into an esophagus if you didn't have to, because they didn't heal well. There was no way I could reach down that far and get hold of anything. I wondered if I could tie her head down to the trailer hitch on the front of the chute and massage it back out through her mouth. We looped the nose tong rope around the hitch and cinched her head down as far as it would go. I knelt down and tried to pinch it back up the esophagus. Again it was painful to the cow and I couldn't move it. Maybe slaughter was the only answer, but she was thin, had a calf on her and wouldn't have much meat on her. We were both quiet for several minutes after releasing her head, then a thought came into my mind. When I had performed rumenotomies to remove metal from cows' stomachs, I'd sometimes poked my fingers into the esophageal opening as a matter of curiosity. If I opened up the stomach, I could push the Kingman tube up the other way and maybe get that thing out. It was apparently too big to pass between the first two ribs, but it might move up the other way.

"What do you think?" His question must have come in response to my expression as I was thinking.

"Well, she could be slaughtered. That's our last resort. I'm wondering if I opened up her stomach and pushed the tube up the other way, if I could push it out through her mouth?" The possibility was there even though it might be a long shot and it entailed an operation opening the stomach.

"Can you cut into her stomach like that?" he asked. He didn't say it, but I knew he meant, "Can you cut into there and still save her?"

"Cutting into the stomach is no problem. I've done that many times when I've taken wire and things out of them. I'm just not sure we can push out whatever's in there." I didn't want to appear to be making any promises.

"Sounds like it's worth a try. How much would that cost?" I could tell that his curiosity had been aroused by the opening of the stomach.

"The rumenotomy would cost thirty-five dollars." I wanted him to know that payment was expected whether we were successful or not. He mulled it over in his mind for awhile.

"I think you'd better go ahead and do it. She's got a calf on her, so she wouldn't have much meat on her if we killed her." He was willing and it looked like I was committed.

"I'll have to get the instruments and everything ready. We'll leave her there in the chute, she'll be all right. I told Nancy to bring the clippers and get some water ready. I began to lay the instruments out in the pan and get the anesthetic and scrub in it. The phone rang and Nancy ran to the office to answer it. In a minute, she re-appeared.

"Dr. Dan, could you talk to this lady?" she asked. "She's got a problem with her dog. I said you were busy, but she's real concerned. I'll finish that and take the clippers out." Nancy was always one jump ahead of me, and I sure appreciated her.

"Hello, this is Dr. Dennis."

"Oh, Dr. Dennis, I'm so worried about Freckles. I guess you can tell this is Mabel. Well you see my neighbor's horses have been coughing a lot and when he came by, I asked him about it. He said they had distemper. The thing that worries me is that when I let Freckles out to go potty, she often goes over into that pasture. I saw her there this morning and called her right home when I thought about those sick horses. I've been worried sick about her, she's surely been exposed. It's been six months since she had any shots. Do you think her shots will protect her? Should I bring her right down? I could come right now."

"Hold on there, Mabel. Don't panic," I replied. "Freckles isn't going to get any distemper from the horses."

"How come? If the horses have distemper and they've been all over that pasture I'm sure the germs are there." I guess I couldn't blame her. A lot of people think dogs get distemper from horses, or like most people they think their horses get it from dogs.

"Well, dogs don't get distemper from horses. The viruses are entirely different. Like a lot of other diseases, they are specific for their own species. They don't infect each other, so Freckles is perfectly safe as far as the horses are concerned."

"Oh, I see. That makes me feel better already."

"Now if she had been around another dog with distemper, it would be a different story. We might think about a new booster shot in that case, but with horses we're home free."

"Gee, Dr. Dennis, I don't know how I can thank you enough. I worry more about Freckles than I did my kids, but she's just like family to me. Thanks again. I guess I'm a real bother to you."

"It's no bother to me, especially when I'm here at the phone and it turns out to be good news like this."

"Thanks, Dr. Dennis, I appreciate you. Bye."

When I got off the phone, Nancy had everything out to the chute. I took out the top board so I could get to the cow's side. I clipped the left side of the cow for the operation. After scrubbing, washing, and painting the area with iodine, I made a hole in the skin with my big needle. The cow didn't like that and tried to kick me. I deadened her through this hole and cut a six-inch incision down through the skin, muscles and into the abdominal cavity. As I was doing all this, I was thinking ahead about pushing whatever was in there out through her mouth. I decided it would probably be a good idea to have the cow's head down a little. I turned to the owner.

"Why don't you catch her with the tongs again and tie here head down a little to that hitch, but not as far down as we had it before." He took the tongs and soon had her head tied and she stopped spraying me with blood from the incision as she moved. "Okay, Nancy. You hold these two forceps and I'll get the other one." When we had all the towel forceps in place, I made an incision in the stomach with my scissors. With the end of the hose cupped in my right hand, I pushed into the stomach to the opening of the esophagus. I entered the opening with my fingers and started the big tube through. I fed the tube into the incision with my left hand and pushed it up the esophagus with my right hand which was in the stomach. I could tell when I got to the obstruction. It was harder to push and the cow began to move and beller in some pain. I could tell the obstruction was moving and once it was past where it had lodged, it didn't bother the cow as much. I guessed it was swollen and sore where it had been. I tried to get a glimpse through the chute and thought I could see that the lump had moved. "I think it's coming. I hope we can get it past her throat."

"Yeah, keep pushing. It's only got a little ways to go." He was excited and was leaning over, kind of twisting like he was helping me to push. I took hold of the tube back about six inches and gave it a good steady hard push. The old cow began to move her mouth in a chewing motion and her tongue kicked out a glob that rolled out on the cement. It was a metal pop can all chewed into a ball with some sharp corners poking out. I could understand why the cow had complained. I sewed up the stomach and the hole in her side. She was good as new in a few days.

Downner Cows

"Dr. Dan, this is Van Winterton. We have a cow down up in Ioka. Could you go up and take a look at her? I don't know that you can do anything for her, but we'd like to know what the trouble is. She might even be dead when you get there." It sounded like he was expecting me to necropsy the cow more than to treat her.

"Yes, I can go right now. I'm not busy at the moment. Where is she at? Do you want me to pick you up?" I asked, realizing he could show me where she was.

"She's in some willows in the upper end. Dad's here, he could go with you. I'm busy in the hay, but he saw where she was. Stop and pick him up here at my house." That was fine with me, I always enjoyed old Mr. Winterton. He was a true pioneer in the cattle business and he had some good stories about his experiences. I checked to see if my axe was in the truck, picked up my necropsy knife and steel, and filled my water tank. I was ready. I stopped at Van's, picked up Hyrum and we were on our way. The trip to Ioka was pleasant enough except I could tell early on that Hyrum didn't think this trip was necessary. In his mind he knew the cow wasn't going to get up, and we wouldn't know any more in the end than we did now. He spent most of our travel time telling me how he had researched the blood lines of the Herefords to eliminate dwarfism from their herd.

We entered the field and he pointed out the willow patch. "She's up in those willows and she isn't ever coming out on her own," he stated. I drove up and parked the truck. He led me back in the willows to the cow. I checked her temperature first. It was normal. Then I felt along her spine and checked the color of her membranes. Everything fell within normal limits and the cow was fairly alert which was usually the case with a downer cow.

"She doesn't have any temperature and I can't feel any injury to the back. Her legs don't appear to be broken. It could be an abscess along the cord."

"That cow will never get up. We should have brought a gun." I could tell he didn't have much confidence in a vet doing her any good, so I didn't discuss any treatment options. The majority of downer cows I'd treated didn't get right up, but it was well worth trying. I went to my truck and picked up some electrolytes and calcium solution. I mixed as much calcium as I could in the electrolyte bottle, popped the needle into the jugular vein and began to run it in. She perked up some and I felt like she was going to respond. When I finished that bottle and started with the calcium solution, her ears came up and she began to shake and quiver. Old Mr. Winterton was standing right in front of her and I knew treated cows sometimes have a chip on their shoulder and want to fight when they do get up. He was well advanced in years and I didn't want that cow to charge him and knock him down.

"You better move back over by the truck," I suggested. "She could be a little on the fight when she gets up."

"She ain't ever going to get up," he said rather smugly. He could see that I was about done, so he did turn and walk a few steps out toward the truck then turned back to watch. I remained as still as I could so I wouldn't startle her. I wanted to let the last of the calcium run down the IV without her moving. She was definitely responding in a positive manner and I felt confident that she was going to get up. As I reached down and pulled out the needle, she threw her head in my direction and jumped right to her feet. There was a moment's hesitation. Then I realized she was zeroing in on Hyrum and he could never get away in time. I jumped out in front of her on the run, waving the empty bottle and IV set I had in my hand. It worked and she turned my direction and charged, head down. I had a couple of steps head start and with that kind of incentive behind me, a yell erupted and I sprinted down the alley in the willows. When I came to an opening leading off to the right, I took it and led her around in a circle through the willows. Before each turn, she would almost get to me and I'd dodge behind another clump of willows. After a bit she finally decided not to turn after me and ran off toward the other cows.

When I got back to the truck Hyrum had a big smile. I think he got a kick out of that merry chase through the willows.

There were some cases of cows going down after an owner had observed them stumbling and weak, especially crossing a ditch the day or two before. These often had small abscesses in the spinal canal and I would recommend slaughtering them as soon as they went down. We at least salvaged the meat. There were a few in which I could detect a broken vertebra. But the most puzzling of all to me were cows that got stuck in the mud or quicksand. There was no evidence of injury before or after they were killed, but they never would get up. Rex Lamb had quite a few get caught in the sand down at Green River and I don't remember any of them ever getting up. I often wondered if the strain didn't scramble their brain somehow.

The most common of the downer cow syndrome were the big dairy cows that went down right after calving. They appeared to be typical hypocalcemia, or milk fever cases that would not respond to the calcium replacement treatment. They would show the normal shivering or muscle twitching, stomach activity, increased alertness and raising of the head and ears, but wouldn't get up. Veterinarians have all treated them with dextrose, electrolytes, vitamins, protein solutions and cortisone. They used to say they threw the book at them, and they still would not get up.

I remember a milk cow out at Brighton's place northeast of town. I treated her each day for three days. She ate, drank and looked like any cow lying down, but she wouldn't get up. After nine days she was still lying in the middle of the corral and since she couldn't get up they left the gate open when the other cows went to pasture.

That day Dee Allred was moving his string of rodeo horses down that road and they swung in the open gate on the run, made a round of the corral, and ran out. The old cow jumped up and ran out with the horses.

It was a frustrating phenomenon to the veterinary profession and many lectures were presented at the various conventions and meetings. I remember Dr. Grant Jensen giving one at our State Veterinary Association meeting. As he concluded his paper he said he would give us his final treatment. Quote: “Take a black cat by the tail, swing it around your head three times and bring it down on the cow’s back and hang on.”

Losses from Bloat

The digestive systems of cattle and sheep with their four stomachs are little mini-factories where billions of living organisms break down the lignins and fibrous parts of plants. The organisms in turn are then digested into proteins, fats and carbohydrates that are absorbed and used by the animal’s body. In this process there are great volumes of gas produced which must be constantly expelled by eructation or belching. The normal “factory” has a lower liquid layer with solid matter floating in and on it and a gas space above. The esophageal opening is the escape valve for the accumulating gasses. Whenever something interferes with the gas expulsion, the animal becomes bloated as we call it. The gas production, by virtue of the ideal conditions of warmth, moisture and culture media or feed, goes on and if it cannot escape, the pressure builds. The stomachs distend to the point where pressure on the diaphragm stops the animal from breathing and it dies. The two most common causes of interference with the eructation process are frothiness in the stomach and a positional change that covers the opening to the esophagus. One other cause is something caught in the esophagus, like a beet top plugging it.

Lush green legumes and grain together, or legumes alone are the primary culprits in the frothy type bloat where the gas is all mixed in the feed and cannot escape. Usually owners and herdsman are aware of this feed consumption and are on the alert to watch for trouble. In spite of this watchfulness, more cattle and sheep probably die from this kind of bloat than all others. The positional bloating where the animal cannot right itself also takes a toll.

One of the most frustrating occurrences in this area happened in the spring when the cows were heavy with calf. They would lay down on the edge of a furrow or ditch bank and end up with their backs a little downhill to the point where they couldn't roll up on to their briskets. It was always amazing to me how little a depression was needed to trap them. The fluid contents of the stomach covers the opening and the gas could not escape. If someone was around to notice the predicament, a little push or a pull with the tractor before it was too late could save them. The sad thing was that owners would find them that way too late. One particularly dangerous set of circumstances occurred when the snow began to melt. The ditch banks being higher with the snow blown off would dry out first while the level part of the field was still wet. This was inviting to the cows and ended in a fatal mistake for some. They would roll into the ditch on their backs and that was the end. I remember one day when Max Patry found five of his cows dead and had to rescue several more.

16
MAN'S BEST FRIENDS

Embarrassment Not Intended

Sometimes in the treatment of animals, the owners inadvertently end up being embarrassed. I got a call from a lady one day who wanted me to treat her dog for a bad sore behind one ear. She said it looked like it had been cut and maybe needed sewing up. When I examined the dog, there was a lot of drainage down the side of the neck and a wound could be observed as the hair was parted. The dog was very friendly and didn't seem to mind my looking so I decided to clip off some of the hair so I could see and treat the wound. As I proceeded to cut the hair, the clippers would clog up and pull when I encountered the exudate matted hair, but the dog was very patient. The further I clipped each way, the cut kept going and

I ended up shaving the hair off all the way around the neck. When I could see the cut all the way around, I knew pretty well what the problem was. I took a pair of hemostatic forceps, reached down into the cut and pulled up an elastic band. As the owner watched, I pulled it all out and off from the dog's head. She was almost speechless for a time. Then the embarrassment and disgust brought a flood of pleas for forgiveness. As the reality of what happened sunk in, she explained that she had seen the kids playing with the dog, putting the elastic around the ears to make them stand up, but never dreamed that an elastic band could cut into an animal like that. She picked up the elastic band in the sink and washed it off to take home and show the kids.

Another time a lady called and was sure her cat had rabies. It wouldn't eat; it was drooling and foaming from the mouth. Although a case of rabies had never been diagnosed in this area, or at the time to my knowledge in the whole state of Utah, anything was possible. I cautioned her on how to cover it with a big towel, then lift it into a box and bring it in. The cat was calm and didn't object to being handled even with my leather gloves. As I was looking it over, there was a short piece of thread hanging out one side of its mouth. It was attached and when I pulled on it the cat obviously felt pain. I carefully opened the mouth and got a glimpse of a needle caught in the throat. I knew then that it didn't have rabies and a general anesthetic would have to be given to get it out. The owner waited while the cat went to sleep, and I was able to remove the needle, which was sticking up through the soft palate. The embarrassing explanation was that she had seen the cat playing with the thread hanging from her pin cushion.

The deep-throated cough of infectious tracheitis, or kennel cough, prompted a goodly number of owners to believe their dog had something caught in its throat. I'd think to myself, "It's just another one of those, and I'd invite them to bring the dog in so we could check and see. I got a call one day about a dog with something in its throat. I didn't ask any questions about how it was acting, just told them to bring it in.

When the dog arrived it wasn't the usual case. It was obviously distressed and clawed at its face with its front paws. It kept its mouth a bit open. When I took hold of its nose and jaw it jerked and screamed, opening its mouth for a second. I thought I saw something white on one side. In a minute or two when it was calm again, I carefully pulled the side of its mouth open and I could see a piece of bone caught on its back teeth. I got my tooth extractors and once again carefully pulled open the side of the mouth. I got the extractors under the lip and grabbed the piece of bone which practically fell out. The owner said, "Why didn't I think to look in his mouth?"

Owners frequently found dogs and cats with fishhooks sticking through their lips, but invariably they couldn't get them out. When they would arrive, I would take my diagonal wire cutting pliers and cut off the barbed end and out it would come. They always said, "Why didn't I think of that?"

Catching a Big One

"Hello, is this the veterinarian's place?" The voice sounded a bit desperate.

"Yes, this is Dr. Dennis."

"My dog has swallowed a fishhook. What should I do?" He wasn't panicky, but there was a real anxiousness in his voice. The time frame was midafternoon like most of the other cases that I had received over the years. I wondered where he was calling from.

"Where are you?" Most of them were at the phone nearest to where they were fishing.

"I'm at the lodge in Rock Creek and my dog swallowed my fishhook. Can you save him, Doctor?" He was almost pleading now.

"Yes, we can usually get them out. Have you cut the leader or did he break it?" Frequently the dog took off and the hook got deeply set.

"I cut him loose. I stood my pole up against a tree and I'd been using cheese on a three-pronged hook. He grabbed that piece of cheese and swallowed it so quickly I couldn't stop him. I did get hold of him and tried a little bit to pull it out, but it wouldn't come, so I cut the leader off. It's still hanging out of his mouth." I could tell the man was really feeling guilty, like it was all his fault.

"That's good. If we have some leader to work with, it's easier to find the hook. Do you know where my clinic is in Roosevelt?"

"No, but these people here at the lodge must know, they recommended you very highly. I'm sure I'll be able to find it. Can I bring him down now?" He was beginning to feel better, the sound of his voice indicated.

"Yes, I'll be waiting. The clinic is easy to find. When you come into town, you'll see the cemetery on your right and it is on the opposite side of the road just west of that service station and bulk plant. There's a sign out front." I looked at my watch and calculated, if it took them an hour, when they would arrive.

"That should be easy to find. I'll come straight away." When they pulled in, the door popped open and a man got out with a small cocker spaniel in his arms. There was a younger man who had done the driving, probably a member of the family.

"Take him right in there on that table." I pointed the way and followed them into the treatment room. The leader was still dangling from the dog's mouth and I was glad to see that.

"We'll have to put him to sleep," I said as I petted the dog and slipped my hand to the inside of the back leg to check the pulse. "How old is he?" I was mentally assessing the anesthetic risks.

"Let's see. He turned five this past April. How long will it take, Doctor?" The dog was in his prime. The pulse was strong and steady, and his tongue showed a normal pink.

"That can really vary a lot. Some are in the throat, others down the esophagus, and some in the stomach. If we have to operate and take it out of the stomach, it takes a while, probably two hours. In fact, we like to keep them overnight."

I didn't want them to think it was going to be too quick, or on the other hand, worry them too much either.

"I was just wondering if we should go back up and break camp and get ready to go home." There was disappointment in the tone of his voice, and I could guess what the rest of the family would feel like.

"Were you planning on going home tonight?" I'd seen this situation ruin a good time fishing before.

"Oh no, we just got out here. We were planning on staying a few days, but with this." He shrugged his shoulders rather helplessly and looked at me. I could read the disappointment he was feeling.

"Gee, don't let this ruin a good fishing trip. Think of it this way. You caught a twenty pounder." I was trying to lighten things up. "If you want you can leave him here, we'll fix him up and you can swing by on your way home and pick him up. Where do you live?" I didn't want them to miss their time in the mountains.

"We live in Kearns, but we love to come out here fishing. Actually it's as much a camping trip in the mountains as it is fishing. The kids love it." I could tell it was a family treat.

"That's great. I think I know the feeling. My family had a fishing trip every summer as far back as I can remember. In fact, my dad and mom used to go with a team and wagon before I was born, and Rock Creek was always Dad's favorite stream." The years and the memories raced briefly through my mind. How wonderful and special those times were for me.

"It's a great place to be, with the stars overhead at night, the sound of the stream, and the wind in the tall pines. You feel so good you forget the rest of the world." It was tempting to ponder and daydream.

"I'd better quit daydreaming and get this guy taken care of. Why don't you wait until I get him to sleep and we have a look. Sometimes I can get lucky and have them out in a short time, and you could take him back with you. He'd probably wake up on the way, and it would save you a trip." It was about an hour's drive to Rock Creek and he should be mostly awake in that length of time. I drew up the anesthetic in a syringe, clipped his leg, and Nancy held off the vein while I swabbed it and inserted the needle. When he was sleeping soundly, I put the mouth gag between his canine teeth to hold his mouth open.

I had made me several small rigid tubes of several lengths from calf pill guns and plastic tubes. They were just the right size to pass down the esophagus. The longest would go clear into the stomach. With one of these and a pen light, I could usually see the hooks and remove them.

With the mouth held open, I fished the leader through a medium length tube and slid it down the dog's throat holding a little tension on the leader to guide me to the hook. It stopped about two inches down the esophagus.

"Okay, Nance, hold his head up so I can see." With the light I could see that two of the prongs were free and inside of the tube end. He was trying to breath and I had to pull on his tongue to let air pass. "Here, Nance, put a finger on this tongue and hold it out so he can breath." I took another look and the one prong was all that was hooked. I knew the barb was caught in the mucosa, but it shouldn't cause a lot of damage. I took a wrap of leader around my finger to hold all the tension I thought the leader could take, then pushed the tube downward, twisting it slightly as I did it. It felt like the hook came loose, so I began withdrawing the tube as I rotated it back and forth a little. Out came the hook held tightly up against the end of the tube. I held it up to show them as they looked through the door.

"You got it," they yelled in unison and their faces turned to all smiles.

"Yep, we were pretty lucky on this one. The leader was strong and held the hook up tight to the end of the tube. There was only one prong caught. Here you can have your hook back. Just don't try to catch such a big one tomorrow," I joked as I handed him the three-pronged hook.

"No thanks. I think I'll keep that one as a reminder." They couldn't thank me enough and extended their appreciation several times as they petted and loved the little sleeping dog.

"Can we really take him with us?" It came out rather hesitatingly. I could read some uneasy concern.

"Oh, I think so. We usually like to keep them until they're awake, but if you can put him on a flat surface and leave him alone he should wake up okay. He should be coming around by the time you get back to camp. He won't sleep long and I'm satisfied that there wasn't much damage done. He'll be a little groggy for most of the night, so just let him sleep it off where it's quiet and he's alone. On your way back while he's waking up don't talk to him or pet him, it makes him try harder when he isn't ready. They flop around more and fight to come to. A dark quiet surrounding is best." They shook their heads in agreement and understanding. "Now, I don't want him to have anything to eat for at least twenty-four hours. After eight in the morning you can give him some water if he wants it. Watch when he drinks to see if it bothers him. If he takes the water okay then he can have a little food tomorrow night. Mix some water or milk in it so that it's sort of a gruel, quite liquid at first. Then keep him on soft food for several days, no bones or hard stuff. You got all that?" I waited as they thought it through.

"Yeah, I think we can do it. The hardest part will be riding herd on the kids."

"Don't forget those cheese-lined hooks," I chided them. "Remember the main things are let him wake up by himself, no water until after eight in the morning, and no food for twenty-four hours. I'll give him a shot of penicillin and if he takes his water and food all right I think you're home free." They paid their bill and carried the little fellow out and put him in the back of the Suburban. He was already starting to respond. They left a lot happier than when they'd arrived, and I felt good too.

That was one of the easy ones, but not all of them turned out to be so simple. Some of them didn't have good leader material and it would break. When it did, I'd try to snare the hook with a wire, or if it was close enough, I'd grasp it with forceps. Some I pushed into the stomach and some were already there. Those caught in the throat I merely twisted out with needle holders. The set of rigid tubes which I made as necessity demanded worked very well and were used many times over the years.

Jealousy, a Costly Sin

One day a lady came into the clinic with a box tied shut with string. She was all bandaged around her neck and one side of her face and one hand also. As she sat the box down, I asked, "What happened to you?"

"Sam attacked me," she said pointing to the box. "I want him put to sleep."

"That old cat did all that to you?"

"He sure did and I can't trust him anymore." I was a bit puzzled because over the last few years she had brought Sam in for shots and check-ups and seemed fond of him.

"How did it happen?"

"Well I guess he got jealous of my little white dog. I've only had him about a week. Yesterday I took the dog to the store with me for groceries and when I came in the door with the groceries in one hand and the dog in the other, Sam leaped up on my shoulder and began to bite me. I dropped the groceries and the dog. I grabbed him to pull him off and he bit my hand. It's time for him to go."

"Well, I guess. I never heard of such a thing."

"I could tell he was upset. When I first got the little dog he spit at him and growled, but I didn't think he'd attack me. He's getting old anyway."

"OK, just leave him and I'll take care of it."

Boarding Dogs

Even before I built the clinic, I used to care for a dog or two when someone had to leave town. I'd keep them at the back of the garage in the bank of kennels I brought with me from Idaho. Then after the clinic was built, I had some outside runs and a few more were boarded. The worst part of boarding was the barking and howling. When a new character is added to the kennels, they begin to voice their opinions, and the whole group chime in and it about drives you and the neighbors up the wall. Most of them settle down within the first day, but there are some that go on for a week. One thing we learned was not to try talking to someone where the dogs could hear you or they joined in.

The pleasure comes with those dogs that are left often, that feel that the kennel is their second home and we are their family, too. Some of them jump out of the car and can't wait to get through the door, while others are reluctant to be left but become happy once the owner leaves.

The fear of one getting away or hurt or sick can be very worrisome. It's agonizing when it happens.

I remember one dog that was brought to board for about three days that could climb a chain link fence like a monkey. My kennel fences were seven feet high with a pipe around the top and I didn't think anything could get out of there. I kept wondering if the receptionist had moved him to different pens when I'd check on him. I tried to keep close tabs on him as I worked out in the corral where I could watch, but I never did see him do any climbing. Then about midafternoon we looked out the front window and there stood that dog looking up and down the highway. We dropped everything and went out to see if we could coax him up to us and capture him. He wouldn't have anything to do with either of us and moved off toward town. I followed him all around town, alternately losing then finding him. When he would run out into the street or the highway in front of a car, my heart would stop. Finally, at dusk, I lost him. I gave up trying to keep track of him and went back to the office. I tried to call the owner just in case someone might be home, and also called a close neighbor, but got no answers. It was a long and troubled night worrying about what could happen. The next morning, I was able to contact the neighbor and to explain my predicament. He told me he had seen the dog over to its home early that morning and told me not to worry that he would go feed and take care of it. I told him how worried I was about having to tell the owner. He just chuckled a little and said not to worry because the owner knew all about the dog's fence-climbing ability. After that one, I bought some more chain link fence and put a lid on one kennel.

Getting away wasn't the only agonizing aspect of boarding. One day we heard a ruckus and went out to find a dog hanging by his choker chain which was caught over the top of the fence. We were glad the dogs had made a racket to let us know.

Trying to keep the kennels clean had its frustrations too. We usually moved all of the dogs over into the clean half of the kennels so we could scrape up the droppings and scrub down the floors with a hose and brush in the dirty half. In eighty percent or more of the cases, those dogs would smell around their new pen and immediately proceed to have a bowel movement. It must have been a throwback to their ancestral habit of marking their territory. As soon as it was done they seemed happy and contented, but we were ready to pull our hair.

Like all other phases of building the practice, the kennels eventually gained a roof and outer boundary fence of cinder block up about four and a half feet. Then over the years came sliding windows to inclose them against the cold, and eventually a separate building to house the pound animals. I didn't have much to invest in these changes except my own free labor, but I was able to work at it from time to time.

Boarding dogs even had a horror aspect to it. One man picked up his dog and tied it in the back of his pickup with a short rope. As he entered the highway and sped across the oncoming lane of traffic to head toward town, his dog decided to jump out. The rope was just long enough for the dog to get over the side and almost to the ground before it tightened, spun him around in midair and threw his head under the back wheel. We were watching it all through the front window but were helpless to even call out until too late.

The highway has always been a threat to all animals that might escape—dogs, cows, horses, and especially to fleeing cats. Frequently, owners carried their cats into the clinic in their arms, or attempted to. After a not too pleasant car ride the cats were frightened and might jump out when the door was opened, or spring free from their owner's arms on the way into the clinic. The other pattern was a cat having just been treated and returned to its owner, sought for more assurance by climbing up around the neck, especially when they started for the car. The claws digging in wasn't too pleasant so the owner would bend down and let the cat jump free thinking they could pick it up again. The cat usually dashed under the car and it became a cat and mouse game in reverse trying to coax them out to get a hold of again. Not infrequently, the cat made a break for it and headed right out across the highway, or when it jumped from their arms it landed running in that direction. The close calls in that speeding traffic have nearly stopped our hearts, but so far none have been hit.

The Two-for-One Deal

I sold Leon Pritchard some land for his crane service yard and got to be good friends with him and his family. One day his wife, Agnes, brought in two cats to be neutered. "Is it cheaper if they come in two at a time?" she said, kind of joking with me.

"I'm afraid not, only by the dozen," I replied with a grin.

"Can I pick them up this afternoon?"

"Any special reason? I usually like to keep them overnight."

"Oh, no I'd just be coming by after work. I can get them tomorrow. It's no big deal."

That evening when I got around to doing the surgery I was alone, so I gave one the anesthesia shot and waited while it went to sleep. Then I gave the other one its shot and proceeded to operate on the first one, thinking it'll already be asleep now when I get this one done. Since they had come in the morning we had put litter pans in the cages with them in the afternoon. When I finished the first one, I took it back to the kennel room and put it in a cage to recover. As I turned to pick up the second one, my heart sank right through the floor. It had crawled into the litter pan and turned sideways with its head out over the side. The anesthesia must have caught up with it in this position, because its head was hanging over the edge of the pan and it wasn't breathing. I grabbed it and gave it mouth to mouth resuscitation and heart massage, but it was dead and didn't respond. I was sick and kicking myself for not taking the litter pan out, but who would have thought the cat would get in that position. It just seemed impossible, but it happened.

I called them up and apologized. I told them how sorry I was and how bad I felt. I tried to explain how it happened, but I didn't think they understood much more than it was dead. I worried all night about having to face them the next morning. When they came in they were sad and a bit cool as they picked up the other cat, but I could tell we were still friends. When Agnes asked me how much she owed I said, "You just happened to get in on our two-for-one sale with the special qualification."

"What's the qualification?"

"I do two for the price of one, but you only get to take one home."

Over the years we've had quite a few laughs over the two-for-one sale.

Sore-Eyed Dog

"Basin Vet Clinic, may I help you?"

"Doctor Dennis, this is Vera. My little dog must have an eye infection."

"What makes you say that?"

"Well, his eye is all mattery. I've washed it out and put murine in it, but it doesn't seem to help."

"How long has it been going on?"

"At least a month. It wasn't bad at first, but it seems to be getting worse."

My first thought about mattery eyes lasting a month made me think about canine distemper.

"How old is he?"

"He's only three months old. I've only had him a little over a month."

"Has he had any shots?"

"No. I didn't think you were supposed to give them any until they were six months old."

"The best time to start them is at eight weeks. But let me ask you, has he been sick or off his feed?"

"He eats like a horse and runs around chewing on everything he can find. He seems healthy to me except for his eye."

"Did you say it was just one eye?"

"Yes. It's his right eye."

"I think you'd better bring him in and let me have a look. We could start his shots too. He's past due on them."

"All right. I was thinking I'd better have him looked at. That eye is beginning to look a little cloudy. Can I come in right now?"

"You sure can. I'm going to be here for awhile."

Later, in the exam room, I tried to look at the eye as he played and wiggled around. When I held his head and pressed the eye closed, I noticed that the eyelid rolled over allowing the lashes to rub on the eyeball. I knew what the problem was. It was entropion.

"Here, if I can hold him still enough, I'll show you what's making his eye sore. Now watch that lower lid when I close it. See that eyelid turn over up next to the corner. It turns over and lets the hair and eyelashes rub on his eyeball."

"Well, I'll be. I'd never noticed that before. I'll bet that feels awful."

"Yes, it does irritate the eye, but there's another thing happening. In the normal eye the lids and eyeball fit together to form a seal as it were. There are little glands in the lid that secrete a waxy lubricant that helps form a seal which keeps everything from getting under the lid. When they close their eyes, the lids wipe the eye off like windshield wipers. When they have entropion it not only irritates the eye, but it allows contamination to get in and the eye is constantly infected."

"Is there anything you can do?"

"Oh, it's correctable, but it takes an operation."

"You mean you have to cut a muscle or something?"

"No, we don't worry about the muscles, although they probably play a part in the problem. We just take out a little half moon of skin next to the lid and that keeps it from turning. We have to judge each one by how bad it turns and how loose the skin is."

"Do very many dogs have this problem? I guess they're born with it, aren't they? I've never heard of it before."

"We don't see it very often in dogs, but there are some breeds that are more prone to have it than others. The interesting thing is that I see it most often in lambs. I probably see ten lambs with it for every dog that comes in. Sometimes on them I have to do both lids, because both upper and lower are turning in."

"Can you operate on him right now?"

"Well they have to be anesthetized to operate, and we use a pre-anesthetic to reduce the amount of medicine it takes, so we like to keep them overnight. I'll have time to do it today if you want to leave him."

"Why don't we. What time can I get him tomorrow?"

"Any time after we open at 8:30. You wanted to start his distemper series, didn't you?"

"Oh yes, I'm glad you reminded me."

"Here, you hold him a minute and I'll get him his shot. Then we can find him a home."

"Will it hurt?"

"Not too much. Most of them don't even act like they feel it. It's just a little prick and it's all over."

"He didn't even act like he knew it was done. He just wants to play."

"Yeah. Most of them are about the same. Here, let me take him and I'll put him in a kennel."

"Thanks, Dr. Dennis. I sure do appreciate knowing what the problem is and I'm glad you're here to fix him up. I'll see you tomorrow."

Puppies, Kittens and Maggots

In the late spring and summer months, I would get calls about baby puppies or kittens with little sores on them. They wanted to know what they could put on them to heal them up. I could almost always count on the originating cause being from fly-blow. The babies climbing around the mother and all over one another would get urine and fecal material on their fur and the flies would lay their eggs there. In a day or so, the eggs hatched into larvae or maggots that began to eat the soiled fur and kept right on burrowing into the tender skin. It was more common than one would think, with newborns coming along when it was warm.

I would suggest that they bring them in so I could determine the cause, and tell them to look over all the others carefully to see if they had any sores. When they were reluctant to bring them in, I would tell them to wipe off the wound and look down into the hole and see if they could see anything moving. They would ask what I meant by something moving. I would suggest that there were maggots in those holes.

Most of the time when they looked the maggots were easy to see and they were quick to bring them in. When they arrived, some of them would be a little green around the gills, so to speak, from just the thought of it. I would take a small pair of forceps and reach down into the small holes and pull out huge squirming maggots, sometimes half a dozen from one hole. The owners always felt so bad thinking they had somehow neglected the poor little critters, and some it made very sick. With the aid of an insecticide to assure that all were removed, the animals would heal in just a day or two if a body cavity hadn't been penetrated. The poor little critters would squirm and complain while I was probing around in there, but how good it made me feel to help out a tiny little helpless animal that was literally being eaten alive.

Just Needs Shots

I was sitting at my desk looking out the window when this car pulled up and a plumpish lady got out. It was Buelah and she reached back in the car and picked up Mitzy. She had called me earlier wanting to bring Mitzy in for her shots. I could see she was talking to the dog as she cradled it in her arms and came into the clinic. Mitzy was a cross between a chauchau and something with short legs. She was rather plump, like her owner, and as spoiled as they come. When she came through the door, the dog began to climb and cling. I motioned for her to take the dog into the exam/treatment room.

"Put her there on the table and I'll get her shots."

"She's scared to death. She knows this place. She's just shaking."

"It sure is amazing how they remember. It only takes one visit and they know the next time as soon as they see the place." I went after the vaccine from the walk-in refrigerator and came back into the room.

"Look at her shake," she said as she put her arms around Mitzy who immediately tried to climb up onto her shoulder.

"I told her we were going to see the doctor to get her shots right after I called you. She didn't even want to get in the car. You know she understands every word I say."

"Set her right on the table and you can hold her head. It'll make her more confident if she can feel your presence." She pushed her back down onto the table and when I reached out to pet her she growled and bared her teeth.

"Mitzy. Now you stop that." I turned back to the sink and got the vaccine in the syringes.

"Maybe if you take a good hold there on the back of her neck and pull her over against you, I can have it done before she knows what's up."

"There, there, Mitzy. Mommy's right here. I'll just hold her head in my hands. I can't watch." She took the dog's head between her hands and turned her head.

I reached out to pull up the skin with my left hand while

holding the syringe in my right one. Before I could even get hold of the skin she jerked her head out of Buelah's hands and dove for my hand with lips pulled back and teeth snapping. With a few years of experience and still responsive reflexes I got my hand out of the way and she continued her dive for the edge of the table. In my minds eye I could see a broken leg or two, if that fat little body crashed to the tile floor. My reflexes once more responded and I grabbed her by the back of the neck just before she became airborne. She immediately began to scream at the top of her lungs and twisted her head around trying to bite my arm and hand as I pushed her back onto the table.

"Oh, don't hurt her. Don't hurt her. Here, Mitzy," Buelah said as she started to reach out to her. By this time I had laid the syringe down and was trying to shore up my grip on the nap of the neck so I wouldn't loose a finger or get bit. I pushed her hands out of reach of the snapping teeth as the dog turned toward her.

"Be careful. She might not care who she bites," I said as pleasantly as I could under the circumstances, and the dog wet all over the table.

"Don't hurt her. Oh you're hurting her." She moaned as urine flew from her twisting back end and the flailing tail.

"I'm not hurting her. She's just scared. If I let her go, she might jump and break a leg. Nancy, hand me the syringes. I'll get the shots over with while I've got a hold on her." My office girl had now joined me. She handed me the distemper shot and began to mop up urine with paper towels. The dog was still screaming, although she was beginning to slow up some, but when I gave her the shot, it accelerated once again. When I tried to give her the rabies shot in the leg, she jumped and twisted around. Nancy grabbed the back leg and pulled it back while I gave the shot and this stimulated the bowels to add their contents. We wiped her off as best we could and I pointed her toward the owner and released my grip. She climbed up the front of Buelah as she began to console her.

"Poor little itsy bitsy, Mitzy. Was that nasty old doctor

mean to you? Did he stick you with those horrible needles. It's all right. We can go home now and I'll give you a chocolate for being a good girl." She held her in her arms and talked to her. "Oh by the way, could we trim her toenails while we're here. She hates to come so bad, I hate to bring her back." I had begun to relax and felt some relief from having completed the job. Now we had to start all over.

"She isn't going to like that any more than the shots."

"No, she probably won't like it, but she hadn't ought to be so scared as she was with those shots. I hate shots. I know she needs it. Maybe she won't make such a fuss over just getting her nails trimmed." My experience didn't agree with that thinking, because most dogs hate nail trimming more than shots. The dog had climbed up until its head and front feet were nearly over the shoulder and it was digging into her neck with its claws.

"Here let me get a hold of her again and maybe Nancy can put a muzzle on her mouth." I slid my hand up over her back and took hold of the skin on the nap of her neck again. She didn't scream, but hung on with all her might.

"Ow, Mitzy, you're hurting me," she screamed and leaned toward me while I tried to get the dog free. I could see that I needed to get my other hand underneath the dog. Now was the time when I felt like I couldn't watch. I had to push my hand up between her breasts to get under the dog. That was terribly embarrassing to me,but I guess the pain of the scratching feet diverted her mind. I lifted Mitzy free and sat her on the table.

"Nancy, get that gauze and let's put a muzzle on her." When she tried to put the loop over her mouth she snapped and shook her head and began to scream again. I tightened my grip and Nancy was finally able to get the loop around the mouth. She pulled hard on the gauze closing the mouth and stifled the screaming. What a relief. We tied another knot under her jaws and pulled it up behind her ears and tied it again. When she couldn't scream she gave up and calmed down some although she kept jerking her legs and squirming as I worked on her toenails. On one of her front legs the nail on the dewclaw had grown so much it was in a circle and growing back into the toe.

"Look at this," I said as I reached for the other style of nail trimmers and cut the nail in two. When I pulled that nail out, there was a mattery oozing hole right into the toe.

"Oh! That makes me sick. I didn't know that could happen. Poor, Mitzy. Mamma's been neglecting you," she baby talked to her. "Will it get better?"

"Yes. Just getting it out of there and stopping the pain is about ninety percent of the cure. We'll swab it with a little iodine and it'll be all better in a few days. I'll bet it feels a lot better already."

"That's why she'd hold that leg up every now and then. I've noticed her holding it up for weeks. Poor, Mitzy, you couldn't tell Mamma when you were hurting."

We finished the nails and pulled off the muzzle. She was either worn out or mending her ways because she edged over against Buelah and acted half normal. I didn't try to pet her and make friends, because I didn't think her ways were that mended.

"I'll go put her in the car and come back and pay you." She gathered her into her arms and began to baby talk to her as she kissed her on the head. "My poor little itsy bitsy, Mitzy, Mamma's neglected you and they've been mean to you. When you get home you can have two chocolates, yes, you can."

Toothache

"It's for you, Doctor," Nancy said as she handed me the phone.

"Hello. This is Dr. Dan," I said as cheerfully as I could having been interrupted while filling out one of the end-of-the-quarter tax reports.

"Doctor Dennis, I think my dog has a toothache." I wondered how she could tell if a dog had a toothache.

"What makes you think it has a toothache? Is the jaw swelled up or something?" I didn't know whether it was a boy or a girl dog, and that voice was a mystery.

"Well, I know her teeth must be getting rotten, because I was playing with her with an old sweater she likes to chew on. I'd hold it and she would have a tug-o-war pulling and throwing her head back and forth. When she jumped to get a new bite, I noticed one of her teeth sticking in the sweater. It must have pulled right out. But the main thing is she doesn't seem to be able to sleep. Every morning early, I'll wake up and she'll be chewing on something like a shoe or a glove. I just put two and two together and figured out she must be having a toothache, with that tooth coming out and her chewing because she can't sleep. What do you think?" I'd heard most of what she said but my mind was frantically trying to put a face and a name on that voice.

"How old is your dog?" One more question, maybe I'd get a clue.

"Oh, you know Gertrude. I call her Gertie for short. I brought her in for her distemper shot when she was a pup." I knew the face, now what was the name?

"Is this Wanda?" I thought that was her name.

"Yes, Wanda Weary, and you remember Gertie don't you?"

"Oh, yes. Now I remember. I couldn't place your voice there for a minute." I'd finally remembered, but what was the problem? Ah, toothache, that was it. "Wanda, I don't think Gertie's got any toothache. That tooth you saw was just one of her baby teeth that she's shedding and that chewing is a pup's second nature."

"She's about grown now. I thought maybe she was hurting because she's been chewing so long, and that tooth coming out scared me. I just started to worry. I've grown so attached to her I don't want anything to go wrong with her."

"How's her appetite?"

"Eats like a horse."

"Does she seem to feel good? Run around and play?"

"Oh, ya. She's active as can be and just loves to play tug-o-war. She's mischievous as the dickens."

"Sounds like the only thing you need to worry about is to

get her in for her booster shots when her year is up."

"Oh, I'll do that. I've got it marked on the calendar. Thanks a lot, Doctor Dennis. I'm sorry to bother you."

"It's no bother, I'm just glad to help." I guess.

Feline Friends

The cat, like man's best friend, the dog, has been with us for as long as history records. It is an amazing animal that can be a little girl's make believe baby to be mauled, lain on, wrapped up, held upside down and even bathed. Then that same night return to its primitive instincts of survival and kill. It is energized with all those instincts of the wild, to stalk its prey, kill to eat, defend its territory, escape its enemies and perpetuate its kind. Man has kept them around for many reasons, but primarily as an aid in controlling mice and rat populations in and around their homes. The importance of the cat is expressed parenthetically in the boast of the rancher who said his ranch was so far out in the hills that he had to have his own tomcat.

When I went to vet school in the late forties, we didn't spend a lot of time studying about felines. We covered some anatomical differences, drug idiosyncrasies, fungal infections, and the common surgical procedures. The birth of the expanding knowledge of cat diseases was just beginning to take place. The cat was starting to take its place as an important pet in the inner-city homes. When I came back here to practice it was still primarily an outside animal more than a pet. I think the facetious comment that a cat has nine lives is a strong tribute to its ability to fend for itself and survive.

As the years passed, more and more cats became loving pets, and I was called upon to treat them more. The territorial battles of the males left bite wounds that abscessed and had to be treated and frequent requests for neutering to prevent it. Then shredded furniture and curtains spawned declawing procedures, and outbreaks of disease required a lot of vaccinations. Certainly the knowledge explosion of the twentieth century pertained to the domestic feline and I was hard pressed to try and keep up. A few selected cases might be interesting.

The Too Efficient Cat

One day I received a phone call that was very puzzling and amusing at the same time.

"Dr. Dennis? Is this Dr. Dennis?" It was a lady's voice, but not one that I recognized.

"Yes, this is Dr. Dan."

"Dr. Dennis, do you have anything that you could give a cat to make it stop catching mice?" There was a short pause as I tried to figure out if this was some kind of a joke, or if she was serious. I finally concluded that it was a sincere request or at least for my part I'd better treat it as such, but why would a person not want a cat to catch mice. I needed more information.

"I thought cats were supposed to catch mice. At least most people hope they will." I was hoping she wasn't one of those people who couldn't stand the thought of a poor mouse being killed.

"Well, that's true," she laughed. "But this dumb cat of mine goes off down in the field and eats six or seven mice, then comes back up on the porch looking like a balloon. In a little while she gets sick and vomits them up all over the porch like she wants to show me how good she is. It's a real mess to clean up, and in a little while, off she goes again. I was just wondering if there was anything we could give her to stop her from catching so many mice?" I began to realize that she did have a problem and my mind raced through the effects of the few drugs cats could take.

"There isn't any medicine that I know of, and if there were you'd have to give it to her every day," I told her. "I can't think of anything that would do what you want or be practical." I thought about the dog or two I'd removed the canine teeth on because they were bad to bite people. "We could remove her front claws, and I have removed the long corner teeth, called the canine teeth, on a dog or two. I don't know that it would stop her, but it might slow her up some."

"Oh, I wouldn't want to do anything like that." Neither did I. The more I thought about that cat, the more I admired it, but it was a problem to her.

"Would you want to just get rid of her? I'll bet there are a lot of people out there who'd love to have her. Why don't you put her on classifieds at the radio station?"

"I couldn't do that. I love her almost like a kid. I raised her from a kitten. I'll bet there are a lot of people that would like to have her." She paused for a moment. "I guess I'll put up with the mess," she sighed. "Thanks a lot, Dr. Dennis. I'm sorry to have bothered you."

Beauty Misleading

I was in the process of preparing Samantha for a spay operation. The owner had brought her in to have her spayed so she wouldn't wander so much at night. She hadn't ever had any kittens, but she was spending a lot of time out at night and the owner didn't want any kittens anyway.

I had everything laid out in the surgery room and had given the cat her anesthetic shot. When she was about fully under anesthesia, I took her from the cage to clip and prep her for the surgery. As I laid her on the table where we prepped them, I took her by the tail and looked to see if she was a female. This was a routine we always hoped we wouldn't forget. When some of the owners brought them in we didn't always check them in their presence, because it might embarrass them to find Tom was Jane. It always seemed easier on the phone. Sure enough Samantha was a tomcat and we had failed to check her until she was asleep for the operation. I'd have to call Mrs. Fuller.

"Hello, Mrs. Fuller?"

"Yes."

"This is Dr. Dennis at the Vet Clinic."

"Is something wrong? Is Samantha all right?" she cut in,

with panic evident in her voice.

"Samantha is fine. She's just gone under the anesthetic and is sleeping, but we have a problem. She isn't a female. She's a tomcat." There were a few moments of silence.

"Are you sure?" The panic was gone, but the unbelief was real as she said it.

"There's no question that Samantha is a male. I was just wondering if you wanted him neutered while he's asleep?"

"Well, I'll be. I can hardly believe that. She was such a pretty cat I just knew she had to be a girl!" There was another period of silence. "She was so beautiful." Another pause. "Maybe that's why she…a…he wanted out so much. I guess if she…a…he is already asleep you'd better go ahead and neuter him."

"Thanks, Mrs Fuller. I'm glad I caught you home."

The A.W.O.L. Cat

It was my privilege at times to hear some amazing accounts of animal's abilities to find their way back home. I've experienced some myself. The following story, one for *Ripley's Believe It Or Not*, was related to me by a family out Ballard way. I wish I could remember who it was, but after many years my memory won't bring it back.

The family raised a kitten in the yard with the old family dog. I think it was a Siamese cross. They tolerated each other and formed a bond of friendship. As the cat reached adulthood and winter came on with below-zero temperatures, they noticed it would curl up on top of the dog in the little cubicle where he slept. That in itself was a rare thing to happen. The cat belonged to the daughter and she and her husband lived with the parents for awhile. Then he enlisted in the Marines and was to report to Quantico, Virginia, for his training. They packed their belongings in the car, took the cat in a small cage and traveled to Virginia. In a day or so they found a place to live and moved in.

Then the parents received a phone call from a sad and

homesick daughter reporting that after a few days she had let the cat out in the backyard and it had disappeared. It never reappeared during the whole basic training period and they moved on with only a memory of their pet.

I think it was the second winter afterward when the temperature was hitting the twenty-below range again. The parents came home from town, and there curled up on top of the old dog was a cat that looked just like the cat that went A.W.O.L. from Quantico, Virginia. Was it him or not? You decide.

Test of Patience

There were many things to test a person's patience and the free wind and water calls, to borrow a line from the service station operators, were often at the top of the list. They expected you to diagnose their problems over the phone without seeing the animal and prescribe a treatment they could buy at the drug or feed store. Some even brought animals for you to see in their arms or vehicles, being careful not to put them on the table because that would constitute an office call. Most of the time I was pleased that they would come to me, but occasionally I'd sure have to bite my tongue. One lady who lived out on a farm augmented their meager income by breeding and selling poodles. She would call or come in to the clinic and bend my ear for long periods of time wanting to know how to treat this or that condition, or how to get dogs to breed, usually just after they went out of heat. The hundred and one things were sometimes amusing and I didn't mind trying to help her. About the only animals she ever brought in were some with lice or mange that one could see from a distance, they were so bad. I think the only thing she ever bought from me was medicine for the parasite problems.

When we started to operate the dog pound for the city she would frequently call and come in to adopt out any little dog she thought she could sell. Like a lot of dog breeders she would give them the first puppy shot when they were about six weeks and then when she sold them she would tell the buyers that they'd had all their shots. She had her problems with mange and distemper both, but we would refer prospective buyers for purebred poodles out to her until one day when she came for a pound dog. It was a poodle mixed with something else. She stood it on the counter and said to Nancy and me, "She's got enough poodle in her if I breed her to my dog, she could have some purebred poodle puppies, don't you think?"

Too Many Dinners

As we were sitting in the office one morning my son Mark, also a veterinarian, answered the phone.

"Basin Veterinary Clinic, may I help you?"

"Dr. Dennis, can you tell if a dog has puppies inside her?"

"Well, yes we can. If we can't feel them we could always take an X-ray. Why do you ask?"

"My dog had her puppies two days ago, but I think she still has one in her."

"How many did she have?"

"She had seven, but I know there is still one in her."

"Did she have trouble having them?"

"No, she didn't seem to."

"Is she straining now like she's trying to have more?"

"No, she's just in there taking care of her babies. She seems normal enough."

"Can you feel another puppy inside her tummy?"

"No, I haven't tried to feel anything."

"She's probably having some discharge. Have you noticed whether it's bloody or green?"

"It looks kind of brown and sticky."

"If she isn't straining, acts normal and has a brownish discharge, it sounds like she's had all her puppies and is getting along fine. Has she eaten anything?"

"Oh yes, she's eating, although she doesn't like to leave her babies, but I know she's got another one to be born." Mark was really puzzled by her insistence that there was another puppy not yet born.

"What makes you think she hasn't had them all if everything seems normal?"

"Well, you see, she was lying out flat with the puppies nursing and I discovered she had this one more dinner, or, a, udder, so I knew she had one more puppy inside of her."

"You found what?"

"Well, you see she has eight, uh, what do you call them—udders. I just call them dinners. Anyway she prepared for eight puppies and she's only had seven, so I know she's got one more to be born." Mark could hardly believe his ears.

"Wait now. The number of active nipples doesn't have anything to do with how many puppies she might have."

"But, Dr. Dennis, I don't believe that. You can see that she prepared for eight puppies."

"Are you married?"

"Yes."

"Do you have any children?"

"Yes, I have two. One is a year old and the other one is three. What's that got to do with the dog?"

"When you had your babies, how many dinners did you have milk in?"

"Well, it was both of them."

"And you only had one baby?" There was a thoughtful silence.

"I guess you're right. I never thought about it that way."

"I'm sure if she had her puppies two days ago and one was still not born she would be in obvious trouble. She'd be straining and sick and off feed. You'd know it for sure. Besides seven puppies is a big litter."

"Okay, Dr. Dennis, you're probably right, but I just thought…well, anyway, thanks a lot. Good-bye."

Strychnine Poisoning

"Dr. Dennis, I think my dog has been poisoned. What should I do?" That announcement always sent up a flag or warning in my brain and put me on the alert.

"What is it doing?" I asked since people seem to think everything is poisoned.

"He's nervous and is panting. When I touched him he jumped and went stiff." These were classical signs of strychnine poisoning, so he was right.

"That sounds right. Get him in here to me as fast as you possibly can. Handle him as carefully as you can and as little as possible. Don't pet him or talk to him, and don't slam any doors or make any kind of loud noises. Just hurry." Time was a critical factor.

The mechanism of strychnine poisoning is essentially the shorting out of the nervous system. As nerve impulses travel up and down the nerve tracts they have to pass through synaptic junctions. These are somewhat like controlling switches. There is a fluid between the ends of the nerves that carries the impulse across, but in the process it undergoes a chemical change. This chemical reaction has to be reversed before another impulse can pass. That's a crude analogy of the protective system in man and animals that preserves order in the nervous system. When strychnine is absorbed into the bloodstream and carried to the synapses, it alters the fluid and allows impulses to travel un-impeded. In electrical terms, it shorts them out. The sensory impulses race up the tracts stimulating the motor nerves which contract the muscles. With all the muscles in a convulsive state, the alternate rhythm of the lungs ceases. The animal doesn't breathe, anoxia hits the brain, the animal passes out and eventually dies from anoxic asphyxiation.

If a dog got a heavy dose of strychnine, it could be dead in twenty minutes, although it didn't happen that fast with most of them. I immediately filled a syringe with a barbiturate anesthetic and got the apomorphine pills out. I needed to be ready. I usually went out to the vehicle with syringe in hand when it pulled in. If the animal was having a severe spasm I would grip the top of the front legs as tight as I could. Sometimes this would allow them to relax and breathe, giving me a chance to give them the anesthetic in the vein in the leg. If they stayed in a convulsion, I would have to inject it directly into the heart through the chest wall. If the convulsions were not too severe I used the apomorphine to make them vomit. Sometimes I dropped a small piece under an eyelid, or mixed it in water and gave it with a syringe under the skin. Within three to five minutes they would vomit everything up and go on with the dry heaves for four or five more times. It really made them sick, but if I could get the poison out of the stomach, the animal would ususally be able to detoxify what was already in his bloodstream and tissue. If I couldn't, then I had to use the anesthetic to paralyze the nerves and stop the impulse transmission. The animal would be kept under anesthetic until it could detoxify the poison. This took from a few hours up to a day or so, necessitating some night vigil on my part.

Powdered strychnine sulphate was the first line of defense, after the owner's rifle, against marauding killer dogs and coyotes that attacked their sheep. The dogs sometimes stalked and stampeded cattle through fences. There were no leash laws and no animal control programs and the result was large numbers of dogs that roamed and attacked. When a flock was attacked and animals killed, the owners frequently laced the dead carcasses with strychnine. When the dogs came back to eat, they were poisoned. It was only simple justice, except that innocent dogs sometimes happened by. Even the owner's own dog would get loose and find the meat tempting sometimes.

I was frequently placed between a rock and a hard place as they say. I'd get a call in the morning to come suture up and treat the mangled survivors of an attack, then that night or the next morning, get a call on a poisoned or bullet-wounded dog. There were times when I knew the dog was the killer, but I could never bring myself to short change or neglect something in my effort to save them. I sometimes wished they would die before they arrived, but if they didn't my inward desire to preserve life always made me try. Until the poison was outlawed by the government it was readily available at most pharmacies and it was abused. Malicious poisonings of a neighbor's dog that barked, bit, or otherwise irritated someone did occur, especially in the towns. Some of the home remedies used were interesting and for the most part effective if used early enough. They would force a tablespoon of salt in some water down a dog or give it ipecac to make it vomit. Then sometimes when they were having convulsions, they would cut off some of the tail to bleed them to get the poison out. It worked to a degree.

Strychnine was also used to poison magpies when they were causing trouble. One day I got a call from Dan Oakes over in Vernal. "Dr. Dennis, this is Dan Oakes. My little dog is acting funny. I'm wondering if it might be poisoned?" For some reason, poison was the number one suspicion in every sick animal, big or little.

"What is it acting like?" I queried trying to determine if it was a true emergency.

"Well, she's nervous, and she's panting. She seems jumpy especially when I touch her." She fit the picture of the early stages of strychnine poisoning. He was half an hour away. I wondered if he could make it on time.

"Sounds like strychnine all right, Dan. You'd better get her over here as fast as you can. Handle her quietly, don't hold her or pet her, or even talk to her. Don't slam any doors or make any loud noises. I hope you make it in time."

"Okay, we're on our way. Good-bye." Sort of instinctively I got the syringe filled and everything ready, when a realization came that I could have met him halfway and cut the time in half. I started for the phone, but concluded he'd have already left. I wondered what he was driving, but decided if I missed him, it would delay treatment even more. I thought I'd better just wait. The dog hadn't sounded too bad. I had time for a lot of thoughts. The phone rang. I wondered if the dog had died.

"Basin Veterinary Clinic."

"Is this Dr. Dennis?" It was a woman's voice not Dan's. I felt relieved.

"Yes, this is Dr. Dan. What can I help you with?" It was a familiar voice, but I couldn't quite put a handle on it.

"Dr. Dennis, this is Mabel. I wanted to ask you if Freckles could catch the flu. My good friend, Florence, stopped in this morning and she has had the flu for a week. She didn't want to give it to me so she stayed across the room, but wouldn't you know it, Freckles had to go right over and jump up on the couch beside her. She reached out and patted her on the head. I scolded Freckles and told her to leave Florence alone. I was trying to get her away from that flu bug, but Florence made a fuss over her and she lay her head right in her lap. I didn't want to insult Florence, so I couldn't do a thing. She would stroke her with one hand and then the other, changing her hankie each time, all the time she talked to me. I didn't know what to do. As soon as she left I gave Freckles a bath, but I'm sure she was already contaminated by then. What should I do? Does she need a penicillin shot?" She sounded really worked up.

"No, Mabel. I don't think you need to worry. Remember I told you about the organisms being specific for each species. Freckles can't get the flu from Florence. There are some diseases common to man and animals like rabies, brucellosis and TB. I don't think Freckles will get the flu though."

"Oh, thanks, Dr. Dennis, that makes me feel so much better. I guess I'm just a worry wart when it comes to Freckles. Thanks again for you time. I'll not keep you longer. Good-bye."

When Dan arrived I met him at the car as I usually did. I could see that the little dog was still alive and wasn't in any convulsion. I tapped her on the back with my fingers and she jumped and stiffened out a little. In a moment she relaxed to her nervous panting. "Boy, I'm glad you made it. She doesn't really seem that bad. Why don't you bring her in and I'll give her something to make her vomit. Handle her as quietly as you can." I stepped back in to prepare the apomorphine and he and his wife followed. While I was shaking the syringe to dissolve the medicine I said. "A little while after you hung up, I thought I could have met you halfway and cut the time in half. I didn't know what you were driving and was afraid I'd miss you so I didn't go."

"I never thought of that. Does it work really fast?"

I gave her the shot of apomorphine under the skin on her neck. "There now, that's going to make you sicker than a dog," I chuckled and they smiled. "If we can get her to vomit I think she'll be all right. She must not have got too much or she could be dead by now. If they get a big dose it can kill them in twenty minutes or less."

"Is that right? I didn't know it could work so fast." We watched her pant away there on the table. I was surprised she didn't try to climb up in his arms. In a little while she began to lick and swallow.

"I'll take that pan now, Nance. I think she's about ready." Nancy had gone out to the back room and brought in one of our large feeding pans. The licking became more frequent and then she began to retch. I held her mouth over the pan by holding the skin on the back of her neck. Dogs never want to do it in the pan and will turn their heads invariably. Sometimes on the big dogs it's a real juggling act to keep the pan under the mouth. In about ten minutes after the stomach was empty and the all-out dry heaves had died down the analgesic effect of the apomorphine took over.

"Okay you can take her home, she'll be all right now." There was thanks expressed, I was paid and they left for home. About a week later we got another phone call.

"Dr. Dennis, this is Mrs. Oakes over to Vernal. Dan is on his way over with our dog again. He left about five minutes ago."

"How bad was the dog?"

"She was about the same as last week."

"Well, that's good. She'll probably be all right until he gets here." I didn't feel the pressure this time and I waited until he arrived without getting the syringe filled. When he carried her in she was stiffened out a little, but by squeezing her legs she soon relaxed.

"I'm glad I got here in time." He was pensive as I fixed the apomorphine. "I just can't figure out why someone would want to poison a little dog like her. She never hurts anybody and she doesn't roam around. In fact I've never seen her leave the yard. They must have thrown something into the yard for her to get. I can't believe someone would do that." He was visibly upset. We went through the routine and when she had quit vomiting and was ready to leave, a nagging thought troubled me. She'd been a little bit worse this time than the last and if they'd do it again they would use more strychnine to make sure. She wouldn't make it in time.

"Dan, I think you better take a syringe and a couple of pills home with you. As soon as you see any symptoms put a pill in that syringe and get some water out of the hot-water tap. Let the hot water run until it's good and warm then stick the needle in it and pull some in the syringe. Hot water is usually fairly sterile. You have to shake it awhile to dissolve it, then inject it under the skin like I did. You might not have time to make it over here next time.

"That's a good idea. I can do that, and it might save her life." He settled the bill and went on his way

It was about a week later when he called to tell me the shot had worked and what had happened.

"I gave her the shot like you told me," he began, "and when she vomited I wondered what she had been fed. I took a stick and looked all through that food. The only thing I could find was some small pieces of hamburger. That's when the light dawned. You see, I've been poisoning magpies all spring. We've been overrun by them. Anyway, I've been putting strychnine in hamburger and putting it on top of a pole I have with a platform on it. It's quite high and I have to use an extension ladder to get it up there. I thought everything that couldn't fly was safe. Then I remembered seeing the dog down by that pole. I guess the magpies scratched some of it off the edges and the dog was picking it up off the ground. I've been poisoning my own dog. I don't know why I'm so dumb. I should have figured it out before."

Porcupine Revenge

The slow, lumbering porcupine was easy prey to the unsuspecting dog, horse or other animal, but his revenge was severe. They were a protected species in a sense, because hunters and stockmen seldom killed them. The unwritten law of the West was that the porcupine was an animal that could be captured by hand and used for food in an emergency out in the forest. We had our share of them and those tens of thousands of quills worn by each one provided a formidable defense.

The points of the quills are made up of a hard, horn-like material with arrow point shaped barbs that stick out from the shaft in all directions, all up and down it. Under the microscope they look like a pine tree with branches that point downward. They are very sharp and penetrate the animals' skin very easily. Then the barbs make them very difficult to pull out, making removal a very painful process. Having treated well over a hundred dogs, I still remain puzzled as to why they attack. Perhaps the instincts of the wild make them see a tasty meal in this slow neighbor, or perhaps it's a matter of territory and who reigns there. I suspect it is the latter. The ones I couldn't understand were the dogs that went back time after time. Some dogs learned a lesson the first time, but others seemed to build more animosity with each succeeding episode.

With some like the horse it was a matter of curiosity or accidental contact, and contact was all that it took. When called upon to remove the quills from horses it was most often from the nose. However, there have been some in the feet and legs. In a few cases lameness resulted as the quills invaded tendon sheaths and nerves. Most all horses are touchy about their nose and more often than not the horses who got quills were the pastured, unbroken or infrequently used ones. Since none of them were very amenable to letting me pull quills out of their nose or even be handled, I devised my own method to get the job done. To avoid the striking hoofs and to limit excessive movement I would put them in a narrow ally with poles behind them. Then with a halter on them I would put my coat, a burlap sack or some other suitable material for a blind under the halter straps so that their eyes were covered. With the eyes covered I could carefully sneak up and grasp a quill with my forceps and he would pull it by jerking back. This worked quite well for all those that stuck out. Occasionally, there would be one imbedded with the point sticking through the inside of the lip. These were located by feeling, and removed by pulling on through if you could do it. This was sometimes a challenge as were the quills in the feet and legs.

Removal in dogs almost always required giving them an anesthetic and it was interesting how many owners were amazed at how fast it worked. My first encounter with quill removal came between my junior and senior years at vet school. I was staying at, and covering Dr. Gold's practice in Salt Lake City. It was about midnight when the doorbell and some vigorous knocking roused me from a deep sleep. I slipped on my pants and shirt and hurried to the door. There was a man with two or three day's growth of whiskers on his face accompanied by two teenage boys. Their clothes showed the wrinkles and grime of a few days out in the mountains. The man's immediate reaction was puzzlement, I guess, at seeing me and not Dr. Gold. It was short lived.

"I'm dreadfully sorry to disturb you at this hour," he apologized. I yawned and shook my head to clear out the cobwebs.

"Oh, that's all right. Come in. I'm covering for Dr. Gold while he's at a vet convention back East." The voice and the face were so familiar, but with that growth of beard it didn't register in my sleepy mind. He stepped through the door with his hand outstretched toward me and the boys followed.

"I'm Richard Evans and these are my two boys. We were backpacking in the high Uintahs and this afternoon our dog found a porcupine. By the time we could get there to pull him off, he had killed it and got himself covered with quills. We hiked out as fast as we could and came straight here. He's suffering terribly." I took hold of his hand as he spoke and I could see and hear the Richard L. Evans of the Mormon Tabernacle Choir broadcast and the "Spoken Word." I shook the boys' hands as appreciation flooded over me for this privilege. His vivid portrayal of life's challenges and his inspired wisdom in how to meet them had lifted me many times over the years.

"I'm Dan Dennis. Go bring him in while I slip on some shoes." I was standing there in my bare feet. The boys went out the door while I went the other way to the bedroom to get my shoes. When I stepped through the door again, there was a big tan boxer dog with a whole head that looked white from the quills sticking in it. There were quills down the front of him, in both front legs and in his feet. I had never encountered anything like this before, but there was no time for second thoughts. He dropped down and began to paw at his face with his front feet and utter a warbling half cry, half whine.

"No, Boy, no." The older son had the leash and jerked it gently to get him back on his feet.

"Could you lift him right up on the table and hold him. I'll soon take away most of his troubles." I pointed to the stainless steel operating table as I retrieved a syringe and needle from a pan on the counter and took a bottle from the cupboard. He was a good-sized boxer so I filled the syringe clear up. They had all lifted him on to the table and were holding, petting and talking to him. I showed the older boy how to hold the vein off for me in about the only place that didn't have quills in it. It was hard and I stuck my hand several times getting hold of his lower leg. The quills when moved would hurt him and he'd jerk a little. The needle didn't bother him and I slipped it in the vein, withdrawing on the plunger to be sure it was in there. I injected the anesthetic slowly, watching the old dog. He relaxed, took a deep breath and slumped down. We rolled him out flat on his side and I continued to give him small amounts and check his reflexes. When he was completely under, I pulled the needle.

"There now, you won't hurt any more." Brother Evans gave him a kindly rub down his side. Then turned to me. "You don't know how much we appreciate this. He was so miserable coming off that mountain and riding back here." They all began pulling the bigger quills with their hands. I got some needle holders and hemostatic forceps out of the cabinet and gave each a pair. I moved around to the head of the table and began pulling quills from the nose and mouth.

"Would you mind bringing that garbage pail over?" I pointed to it as I spoke to the youngest lad.

"I want to keep them and count 'em," he said as he looked first at me and then his father for approval. He had a small pile on the corner of the table.

"All right, let's do it." His father could read some special meaning in the request and joined in with enthusiasm. It changed the atmosphere from concentrated total effort with near exhaustion to a measure of fun again. I found him a small box and he counted the quills as we pulled and gave them to him. I don't remember the final number he came up with but I do remember we passed the two thousand mark as I was finishing up his mouth. That dog had quills inside of his mouth, through his tongue, almost down his throat. They were in his eyelids, all the way down his front, on both front legs and his toes were like pin cushions. I've never seen one that bad since.

It was three A.M. when I finished rubbing and feeling each area for hidden quills. The dog was beginning to come around when I said they could take him home. The kind words of appreciation and apology for the late hour were repeated periodically as we worked and then with warmth, sincerity and deep emotion as they prepared to leave.

"How much do I owe you?" he asked. I had no idea. I knew what an office call was supposed to be, but not anything like this.

"I don't know. I'll have to let Dr. Gold send you a bill next week. I already felt paid by the feelings expressed and the satisfaction of helping that dog. It was a special night to remember, and ever after when I would hear or see Richard L. Evans on the choir broadcast, it would come to mind.

The Indian Method

Guy Pinnicoose, a good Indian friend, who lived across the red bridge on the Duchesne River above Randlett, called one night as it was getting dark. “My two dogs found a porcupine. Got quills bad. Can you fix ’em if I come up?” I always enjoyed working for Guy.

“Sure I can fix ’em. Bring them up.”

“Where? To house?” He knew I was home.

“No. Go to the clinic. I’ll meet you there.” I finished my supper and went down to get ready. He pulled in and turned off his lights. I opened the door as he approached with one of the dogs. “Take him in there and put him on that table.” I pointed the way. I gave him the anesthetic in the vein and went to work pulling quills. They were confined to his mouth and nose so Guy just leaned against the wall and watched pretty much in silence.

When I was nearing the end he said, “This a lot easier. I tried Indian method, but too many quills.” I hadn’t ever heard of any Indian method so I was curious.

“What’s the Indian method?” I asked him.

“Put rope around neck, hang over clothesline pole. When pass out, let down, pull fast.” He paused then went on. “This time too many quills.” I was thinking about what he’d told me and mentally comparing it with stories I’d been told about trying to tie and hold dogs to pull quills out of them.

“Sounds like it might work.” It was a rather drastic approach I thought, but I never liked to belittle their native ways. I finished with the dog and laid my forceps in the sink. “Okay, you can put him in your truck and bring in the other one.”

“Nope,” he said with a touch of sadness. “Him didn’t make it. Indian method not too good.” I could tell he wasn’t happy so I didn’t make any comments.

From that time on when I would get calls wanting to know if cutting off the quills or putting something on them would make them release, or wanting me to tell them how to get them out, I was always tempted and occassionally in jest did suggest the Indian method.

Seven-Year Itch

"Come on in here, you two," the woman scolded the two boys. I was a little puzzled. Most of the time when a woman and four kids come along the kids swarm in, climb on every thing, look at every thing and run in and out of every room. Here were two that didn't want to come in.

"Dr. Dennis, can you check our dog for mange?" She didn't have any dog and I hadn't seen any in the car when they all got out.

"Have you got him with you?"

"No, I've just come from the doctor's office and he thinks my kids have this itch, or mange, from our dog, or some dog." There was real concern in her voice; a tension weighed heavily upon her, I could tell. I knew what the problem was and it wasn't that serious. I thought I'd lighten things up a bit.

"So you've got the seven-year itch," I smiled.

"Yes. He did say something about that. He said it comes from dogs. He said if we could find this certain kind of a bug, or I think he called it a mite, on our dog then we'd know for sure if that's what the kids have. Is that true?" She was a little more relaxed but her feelings had shifted to embarrassment.

"It's true. The sarcoptic mange mite of dogs can be contracted by man. It's a little microscopic mite that burrows down into the skin. Here let me show you a picture or drawing of it in this book." I turned and brought out a textbook of parasitology in man and animals that I'd had in school. When I found the page and showed her the drawing of the mite down in its little tunnel and the enlarged picture of the adult mite, an involuntary shudder shook her whole body.

"How awful. Can you get rid of it?" She shuddered again

and scratched at her shoulder.

"Yes. It's quite treatable with the insecticides we have now. It's sometimes a little hard to diagnose. We have to do a skin scraping and look at it under the microscope. With the mite burrowed down in the outer layer of the skin we have to use a sharp blade and scrape down to where it bleeds." I showed her the drawing again.

"Yes. The doctor said he could do that with the kids, but if we could find it on the dog we wouldn't have to. The kids have been scratching for a month or more and I found these little red spots on them. You boys come over here and show Dr. Dennis." The two boys reluctantly came over and pulled up their shirts. There were the little red spots like I'd seen on others before, but these little guys had a lot more. I understood now why they hadn't wanted to come in. "I've tried calamine lotion and alcohol and Mentholatum and everything. They've just been itching like crazy. Why do they call it the seven-year itch?" I could see the two boys secretly scratching themselves as they went back and sat down.

"I guess in earlier times they didn't have insecticides like we do now and the grease and other medicines they used wouldn't penetrate down into the burrows where the mites were to kill them. It must have been pretty miserable. Somewhere along the line it got the name of the seven-year itch." She shuddered again as she held the baby.

"I think I understand that a little bit. Can you cure the dog?" She was definitely ready to get rid of it.

"Yes. We have to give them a medicated bath every week for several treatments, but it's quite successful. Dogs have another kind of a mite called a demodectic mite and it's very difficult to treat. Fortunately it doesn't affect man.

"I hope this stuff the doctor prescribed works on the kids. Maybe I should get some of your medicine and give the kids a bath. That sounds like the easiest way."

"No. We couldn't do that. The medicine we use on the dogs is probably a lot stronger, and it's not cleared for human use. Use what you've got and have a little patience. It takes a little time for the medicine to soak down into where the mites are, but I'm sure it will clear them up. You're not the first one who's been here with mite problems."

"Is that right? I thought I must be an awful mother. It about gave me a complex watching those kids scratch."

"Well, don't feel bad. It happens to the best of us, and it'll all be over in a week or two. Go get the dog and bring him in."

"Okay, kids. Come on, let's go. Thanks, Dr. Dennis. I feel much better. I'll be back in about half an hour with the dog."

Half an hour later she was back with a string around the dog's neck, trying to keep it at least at arm's length. If it brushed her leg, she'd step away as quickly as she could. I picked the dog up and put it on the table and she acted like she was glad to get rid of it. The poor little guy had bare spots scratched all over his body with serum and blood scales hanging in what hair was left. I took a skin scraping in a couple of areas and took it in to the microscope. There was a mite moving under the lens in the first field I looked at. I found one out in the open and called the lady in to have a look. When she bent over and looked in that microscope her breath caught.

"Ow! How horrible. It's moving its legs." She drew back with a shudder, then looked down close with her naked eye. "Where is it?" This was a reaction a lot of people have when they look in a microscope.

"It's there and it has some friends too, but you can't see them without magnification. That poor little dog has really got a problem. I think you better leave him here and we'll treat him now and again tomorrow afternoon. Then you can take him home." We gave the little dog his medicated bath and I noticed Nancy using plenty of soap and a lot of scrubbing when she washed her hands. Just to think about it made you itch and scratch.

17
REFLECTIONS

The Reward of Appreciation

I think the most rewarding aspect of bringing veterinary medicine to the Uintah Basin was the appreciation that was expressed and felt by the people I helped. However, the times that touched my heart most deeply were when animals showed obvious relief or joy following my efforts. Elsewhere, I mentioned the small, thirsty, blind, forage-poisoned steer that kept his head pressed against my hip as I walked him out of the hot field. I had poured water into his mouth to give him a drink and he acted as if I was his total salvation. He didn't want to lose track of help when he was so thirsty.

Another example was the way young heifers would relax and even sleep after their calves were taken from them by caesarean section. While I was suturing up the uterus and abdominal wall many of them would even snore as they relaxed.

The calves that gave me a continuing satisfaction were the chronic bloaters. This is a condition of unknown cause that interferes with the belching reflex. The animals are actually starving to death very slowly. They eat a small amount of feed and the fermentation quickly fills the stomach with gas. They don't get enough eaten to cause the pressures of death, but go around distended with gas like a toy balloon. Usually, by the time they were brought to me they were in poor flesh and weak. I would make a small permanent hole in the top of the first stomach by sewing the skin to the stomach wall. This provided an escape for the gas and they could eat all they wanted. It was always a pleasure for me to see them standing at the manger scarfing down that feed. When the stomach got full, ingesta would run out the hole and they would keep right on eating for a few days until their system was satisfied.

Once in awhile, you have an animal that will actually express their appreciation like the big black Labrador in Idaho and Less Gardner's dog. I went up to Less's place to treat some cows and look at a horse. When we finished, Less told me his old dog had something wrong with his ear. I knew the most common problem was a grass awn down deep in the ear canal and I didn't have the equipment necessary to get it out. I explained that he might have to bring him in to the clinic, but we would take a look. We knelt down on the grass and Less called the old dog over to him and held his head in his big old hands. Less had about the biggest hands I ever saw on man. I looked inside and could see a big tick all engorged with blood. I went to my truck and returned with some long shank hemostatic forceps. Less held him again and the old dog held as still as a cedar post while I reached down in to pull out that tick. I laid it on my hand and held it out to the old dog. He took one little sniff then took it between his teeth and crushed it and flung it away as he shook his head. He seemed to get revenge. He shook his head a few more times then began to wag his tail and we were all happy. Two days later I had to go back up to Less's place and when I got out of the truck that old dog came up to me and rubbed his head against my leg and warbled out a sound of pure joy. I really felt like he was telling me thanks once again.

I certainly felt appreciated and needed by the people as the years slipped by. I remember when I was in the hospital for surgery on my hernias, there were people that came to me about their animals. Later when I had kidney stones and was being wheeled out to be shipped to Salt Lake, one man followed along asking my advice and explaining his problems. The nurse thought it was terrible that he would bother me at a time like that, but I was glad he felt close enough to come to me.

Public Service

I had a great love for my country and the developments following World War II were a great disappointment to me. As a serviceman I thought our victory would bring peace and security, but the Communist threat became more sinister than Hitler and Germany. I also felt our political leaders had betrayed us to an extent with the international agreements. At the same time the creeping socialism in our own country was destroying self initiative and saddling the individual with confiscatory taxation. It really bothered me that every decision a person made was weighted more heavily by the tax consequences than any other factor. To me it was destroying the American way of life and gradually eroding my freedoms. The tax effect on decisions and the regulatory mandates on the individuals and businesses were eliminating free agency and our ability to control our destiny.

For these and other reasons I involved myself in some of the political processes and talked about our problems with those around me. It was at this time that I was having the honor of serving in leadership roles with the Cattleman's Organization and the State Veterinary Medical Association. I found that my opinions and positions on issues were respected and appreciated. My whole life had been blessed with a measure of discernment and foresight.

I guess because of the difficulty of getting candidates willing to serve and my activity, I was approached about running for a position on the city council. I accepted and was elected to that body with the feeling that I could carry on my practice and at the same time do what was needed for the city. The meetings were at night and the consultations and inspections were at odd times that I could make.

The real challenge came when I was approached to run for the legislature. I told them I didn't see how I could do that. It would mean being away from my practice all week with only the weekend to catch up. They argued that it would only be for sixty days every other year. My experience had taught me that I did have some qualifications, and my thoughts kept insisting that one man can make a difference. I had strong feelings that something needed to be done, but I wasn't sure how much could be done to correct our problems on a state level. I began asking my clients how they felt about it and what I should do. They were about as troubled as I was. Some didn't want me to be gone when they might need me. Others really wanted me to run even though it might mean I'd be gone when they needed help. The large animal clients looked at the time period from early January to about the tenth of March and felt there wouldn't be serious need until calving started. I looked at the records and found January to be a very slow month with February about the same. It consisted mostly of regulatory work and the sale on Saturday which I could be home for. A lot of the small animal work could be scheduled for the weekend. Ultimately I made the decision to file and run for one term.

I was a poor campaigner because I didn't like to toot my own horn as it were, and the heat of the campaign came in the fall when I was swamped with work. My opponent, Mr. Cliff Memmott, was an incumbent and a fine member of the community and he owned the local paper which he used most effectively. In spite of optimistic predictions I came in a close second and lost the race. Two years later the district boundaries had been changed to include Wasatch County and Cliff chose not to run, accepting an appointment by the governor to the State Highway Department. This time I was a little wiser and was elected to the House of Representatives for the state of Utah.

When the session started in January, 1967, I worked hard at learning the procedures and burned a lot of midnight oil studying the bills. I wanted to know what each bill would do and how it would affect the people of the state immediately and in the future. I didn't speak on many bills, but the questions that I raised caused many to think and sometimes change their vote or amend the bills. I earned the reputation of being one of the best informed members of the House. The novelty of being the first veterinarian to serve in the legislature helped to make friends. I rapidly gained a lot of respect and knew each member by their first names both in the House and the Senate. The real burden fell on my poor wife. She was alone with a big family and constantly being called about medicine or a problem with an animal. As the session adjourned for the day I would call home to hear how things were going. She would keep a list of their questions and problems and give me their phone numbers. I could pass on advice as needed and make plans for the weekend. I had a phone at my desk and when something urgent came up, I sometimes received calls while the session was going on.

Besides being a solo practitioner I was bishop of the Roosevelt Second Ward while serving in the legislature. Needless to say I was kept busy and provided with enough stress to keep me alert and life interesting. Having to give up two months of earnings put us behind the eight ball financially and we had to consider the legislature as our vacation. Joyce did get to join me for a couple of nice dinners and an activity or so. My legislative salary and perdiem just about paid for my room and meals. If I was careful I could break even.

Before I left for my first session I was charged by the local community leaders with the responsibility to obtain funding for a college program in our area. There were a group of people who had been working on this for a number of years and they wanted me to keep it before the legislature. When I mentioned it to some of my friends and acquaintances in the legislature they told me if I did, I would become the laughing stock of the legislature. I had gone there with a mandate I couldn't ignore, so I consulted state agencies concerned with education for advice on how much it would take and how best to administer it. I introduced a bill and the course and detail of what occurred would constitute a book in itself. By what seemed like a miracle the bill did pass and a program was started. I returned home a local hero and as a highly respected legislator by my colleagues in the House and Senate.

The rather startling success of that first term projected me into serving five more for a total of twelve years. It was a great sacrifice in terms of money and time, while at the same time it brought rich personal rewards. To associate closely with the high calibre of people elected to the legislature was a great and uplifting experience. I gained a much greater respect and confidence in our elective process and its innate selectivity of character and integrity. It works and although in the twelve years there were four or five that I felt were lemons, the majority were of outstanding character. I myself also received an education in studying the state statutes and proposed legislation. It was easily the equivalent of a four-year degree in college. This combined with my association with all the state agencies and higher education in particular gave me choice experiences and much insight. I've always considered it a very choice and outstanding experience.

As much as I enjoyed my service, I still had a great concern for my practice and the welfare of the animals and my clients. My thanks goes out to members of the State Veterinary Medical Association who voluntarily gave of their time to cover for me while I served. I think the thing that hurt the most was the casual remark by friends and colleagues that I would rather be a politician than a vet. I wanted to serve the people of my area and the state of Utah to the best of my ability, while at the same time be the best example I could of the great veterinary profession.

In the spirit of continuing public service following the legislature, I served a term on the Alumni Council of the Utah State University and the bigger part of a term on the State Library Board, which I interrupted to go on a mission for my church.